# Biology Is Outdoors!

## A Comprehensive Resource for Studying School Environments

Judith M. Hancock

illustrated by Rhonda L. Hodges Cline

 **J. Weston Walch, Publisher**
Portland, Maine

# Users' Guide
## to
## *Walch Reproducible Books*

As part of our general effort to provide educational materials which are as practical and economical as possible, we have designated this publication a "reproducible book." The designation means that purchase of the book includes purchase of the right to limited reproduction of all pages on which this symbol appears:

Here is the basic Walch policy: We grant to individual purchasers of this book the right to make sufficient copies of reproducible pages for use by all students of a single teacher. This permission is limited to a single teacher, and does not apply to entire schools or school systems, so institutions purchasing the book should pass the permission on to a single teacher. Copying of the book or its parts for resale is prohibited.

Any questions regarding this policy or requests to purchase further reproduction rights should be addressed to:

Permissions Editor
J. Weston Walch, Publisher
P.O. Box 658
Portland, ME 04104-0658

—*J. Weston Walch, Publisher*

*ACKNOWLEDGMENT*

Many thanks to Rhonda L. Hodges Cline,
who drew the illustrations

1    2    3    4    5    6    7    8    9    10

*DEDICATION*

To
biology teachers and students
for greater understanding of the familiar

# Contents

# Introduction

During most of human history, naturalists have studied the living world in the field, with occasional forays to the lab, but in the 20th century, the laboratory has become the scene of biological activity, and most biologists are lab-oriented. Laboratories and their equipment have become increasingly more complex; the findings ever more refined, abstract, and detached from the natural world. Now among professional biologists, only ecologists retain the old outdoor connection, and even many of their studies are done in the lab. The scene has shifted: Now most naturalists are amateurs who enjoy learning about nature as an avocation or form of recreation.

The increasing sophistication of the research creates problems for biology teachers who struggle to keep reasonably up-to-date and who wonder how much of the new and how much of the old they can transmit to their students. Too often youngsters' sense of wonder at the natural world, and curiosity about it, gets lost in the shuffle as events at the molecular level take center stage in biology classes. For many students, molecular biology is not simply incomprehensible; it is positively sapping to the spirit of wonder and curiosity. It is time to return to nature.

Some teachers continue to make use of that reputable old standby, the field trip, but this is often fraught with problems. Teachers are ill-prepared for field work, since their own education emphasized lab studies. The host of scheduling, transportation, and site selection difficulties can discourage the staunchest field-trip advocate. Besides, lurking in the subconscious is a nagging feeling that somehow such an activity really isn't scientific enough. But field study is valid, and it remains the best way to learn about nature and to stay in touch with it.

This book offers, in place of the conventional field trip, investigations into the familiar world of the school grounds. May you find it helpful.

## *The School Grounds as an Ecosystem*

A school environment is so obviously an artificial environment that you may be tempted to dismiss it as a subject for study. Or you may think that only suburban schools, well endowed with grounds, could provide a suitable environment. In fact, though, life is irrepressible and found in any environment, even the most pavement-surrounded urban school. It is a matter of looking for it.

Any ecosystem comprises both the physical aspect of the environment (climate, geology, geography) and the interacting organisms. In an artificial environment, the terrain has been altered for human purposes, accompanied by the loss of most of the native plants and animals (cutting forests to make fields for agricultural purposes and covering open areas with buildings and pavement are the most obvious examples). After such radical alterations, new plants, whether crops or landscaping, are added, further changing the environment for the animals. New animal species, able to live in the new conditions, move in and become established. Before long, a thriving, self-contained community of life exists in this artificial environment. Such a community is well worth investigating.

Unlike the conventional field trip to distant parts, the school grounds are as near as the door. You don't have to worry about procuring extra time, buses, or personnel. Insurance, which concerns administrators, does not constitute a problem, since the school's insurance covers its grounds. Thus the school grounds provide an unusual educational opportunity: they solve organizational problems while providing a unique ecosystem to study.

## *The Plan of the Book*

Science, a way of learning about the world through objective, repeatable investigation, is always on the cutting edge of knowledge. Science is inquiry. That is one of the most important things for students to understand, and inquiry should be one of the major objectives of any science course. While students may not be on the cutting edge of knowledge—though occasionally they are—they should have the opportunity to investigate real topics for which there are no "right answers" in the textbook.

It is in this spirit, and with the conviction that investigation is the best way for students really to understand science, that this book was undertaken. All the investigations are open-ended and adaptable to different locations; portions can be selected to fit your time and educational goals.

The book consists of ten investigations:

1. Physical Setting of the School
2. Plant Life on the School Grounds
3. Health of the School Grounds' Plants
4. Soil Analysis
5. Soil Organisms
6. Opportunistic Species
7. Microenvironments
8. Impact of the School Building on the Environment
9. Impact of People on the Environment
10. Natural Areas

The investigations are organized into the following sections:

### Teacher's Section

The opening section, addressed to you, contains material that will help you handle that particular investigation. It begins with background information on the subject for your benefit which you can use to give students preliminary information before they begin the investigation.

Following that is a part titled "Procedure" that explains what sorts of things are to be investigated and how to go about doing so, as well as management suggestions. Needed equipment and its use is also explained in this part, though one investigation (5) has a separate section devoted to equipment.

Lists of supplies for the investigation and pertinent references are found at the end of each Teacher's Section.

In addition, the parts titled "Results" and "Discussion" of some Teacher's Sections help you make full educational use of the investigation. Discussion questions focusing on important aspects of the investigation are included with "Comments" to help you lead class discussion. Investigations without "Comments" have answers that will vary according to procedure and location. All the questions, however, do require students to draw thoughtful conclusions from their data.

## Student's Section

This section is written for the student and can be reproduced for your class. It consists primarily of "Directions for Obtaining the Data," a part that closely parallels "Procedure" in the Teacher's Section. There are some differences in phraseology, but the two parts are substantively the same.

The Student's Section also contains Data Sheets. These pages, intended as a common form for recording the investigation's results, are also reproducible.

Finally, suggestions for further investigation called "Spinoff Ideas" are included with some of the investigations (4, 5, 6, 7, 8, 10). They are addressed to the student, can be reproduced, and offer some ideas for **independent research projects.**

# *Educational Aspects of Investigation*

Student investigation is undertaken for one purpose only: its educational benefit. It exists for students, not for the advancement of scientific knowledge. As such, it must provide students with opportunities to pose questions, to devise logical means of answering them, to draw conclusions, and above all to recognize errors they made in procedure or rationale. These are worthwhile skills which will serve all students, whatever paths they choose in life.

In addition, in our scientific and technological society, students should understand how scientific knowledge is acquired. There is no better way for them to do so than by conducting investigations themselves.

In order to insure that education really does happen, and to the best possible extent, you need to be aware of and take into consideration a number of aspects as you begin this venture.

## Site

Field studies are always more difficult to manage than ones in the lab; that's one of the reasons teachers tend to avoid them. For youngsters, outdoors means freedom from the confines, physical and mental, of the classroom. It's recess. To get them to approach outdoor study in the proper frame of mind requires preparation and organization on your part.

## Planning

After you have decided on the amount of time you wish to devote to this subject, you must familiarize yourself with the area in order to see which particular investigations you can use and how to adapt them to your school's grounds. Of necessity, the wording in this book is general, applicable to the wide variety of school grounds found across the country. This book provides the guidelines, but you have to make them specific for your own situation.

Part of your familiarization must include the organisms. You need not pursue this in depth—identifying species, for example—but you do need to know locations where organisms can be found. The emphasis in these investigations is on higher plants and small organisms, and the latter, though omnipresent, are not easy to discover. It helps if you know which bushes are infested with insects or lichens, where good puddles are found, areas of human impact, and little pockets of nature. That knowledge enables you to plan activities most effectively.

## Organization

Organization is of prime importance. You can get away with some deficiencies in planning and familiarity with the area as long as the activity as a whole is organized. The tighter the organization, the more smoothly the investigation will proceed and the better the educational value.

Decide ahead of time how students are going to work—whether individually, in partnerships, in teams, or as a whole class; suggestions are included in the Teacher's Section. In most cases, the investigations are conducted by teams of three or four students. It is preferable for you to assign students to teams. Left to themselves, students will choose to work with their friends, a situation that tends to promote the carnival atmosphere, as well as cliquishness, that you wish to discourage. If you make the assignments, you can avoid team homogeneity, and by changing the teams with each investigation, you force new associations, some of which may develop into friendships. Your selections are also more objective, and the shy or new or different youngster is fully included.

Prepare your students for this activity ahead of time by giving them information on the subject in a way that excites their interest. Your approach makes all the difference: If they catch a feeling of excitement, expectation, and adventure from you, they will be in the right mood for investigation, but if you are flat, disinterested, going through the motions, or apprehensive, your students will not take the activity seriously and will give you all sorts of problems. Other suggestions include giving students a chance to participate in planning, taking their ideas seriously, and being involved yourself in the outdoor activities.

Assign each team its tasks before you go out. Try to plan things so that the teams have about the same work load and that they are busy. One of the best ways to create the working atmosphere that investigation requires is to prevent idleness. Students should be somewhat too busy—not overwhelmed but pushed so they have to keep going. Timing an activity works wonders. So does knowing exactly where to go, what is to be done, and having all the necessary supplies on hand.

When outdoors, monitor activities carefully and be prepared to remove anyone who insists on horsing around, not taking part properly and preventing others from doing so. At the same time, encouraging the strugglers and commenting publicly on work especially well done makes the individual feel noticed and tells the class that you are on top of things. That helps everyone's attitude.

Once the data or materials have been collected outdoors, get your class back to the lab. Don't be persuaded to have a discussion outdoors.

## Objectivity

You do want your students to be honest and objective about their procedures and the accuracy of their results. You also want them to enjoy investigating. Once again, your attitude is the key. Your students cannot be as objective or accurate as you might wish— they are only learners, after all. So while you should stress those desirable qualities, you cannot emphasize results, especially "good" ones, too much, or grade on them, since that tends to encourage students to achieve what they think you want, often at the cost of honesty. Subtlety is called for. Reward the effort more than the results.

## Data Sheets

Most students simply have no idea of what information is important to an investigation or how to record it. With that in mind, the data sheets were set up to give students a form for recording their results. This has several advantages. With everyone using the same form, the data are more consistent and comparisons easier to make. It is also easier to evaluate.

The term "data sheets" was chosen deliberately. These are forms for recording data, the same sort used by professional scientists for their data, not the usual workbook fill-in pages. These are real data your students are recording, and as such they should be recorded at the time observations or experiments are done, which will mostly be outdoors. Data sheets are results of the investigation and should be carefully preserved. You need to emphasize these points to your students.

Students should pay attention to the form of the data sheets, since the sheets teach recording skills—something that is useful not only in science courses but also in other contexts where one must record information. To give students the opportunity to devise a recording form, one investigation (9) that is particularly variable does not have a data sheet.

## Discussion

In scientific investigations, drawing conclusions and seeing their significance is of great importance; indeed, it is the reason for the investigation in the first place. Scientists typically share their results and their thinking with one another by means of papers and seminars.

This is also true for students. The written report serves a valuable purpose, but too often it is seen only by the teacher and the student who wrote it. Sharing ideas with one another is important, for in doing so a student learns about skepticism, disagreement, conclusions that can't be substantiated, nonconforming results, and intellectual give and take. These worthwhile lessons, important aspects of investigation (and life), are best fostered through discussion.

In most cases, individual students do, and hence focus on, only a part of the investigation. Sharing results enables them to see the whole picture and therefore be better able to understand its significance.

Handling a discussion is a supreme exercise in the teaching art. You have to challenge, encourage, referee, restrain, adjudicate, coach, and stimulate while resisting the urge to leap in and provide the "right" conclusion. A tightrope act if ever there was one! Yet if you manage well, your students will become so absorbed that they'll forget everything else, including you, and the discussion will carry on by itself—often even after class. It's very satisfying to see students startled by the ringing bell!

The purpose of the discussion is to share ideas, not to reach consensus, so don't attempt to have students reach general agreement, and be sure they don't feel uncomfortable about unresolved questions.

## Evaluation

Grading of investigations should take all facets into account: performance, data sheets, discussion, and the subjective qualities of attitude, effort, and the attempts to understand and express budding ideas.

# General Supplies

Supplies are listed at the end of each investigation. This list includes all supplies. Numbers in parentheses indicate in which investigation each item is used.

## *Measuring Equipment*

carpenter's steel tape (20–30′ long or several 8′ tapes) (1, 2, 8, 10)

yardsticks (1, 2, 10)

rulers (6″ and 12″) (1, 2, 5, 7, 8, 9)

compasses (1)

level (1)

teaspoons (4, 7, 10)

thermometers

  soil (8, 10)

  weather (7, 9)

PVC coupling or tubing (4″ diameter, about 6″ long (4)

watches (4, 10)

humidity meter (or drying crystals or cold glass) (9)

calibrated line (10)

Secchi disk (10)

marker flags (10)

diet scales (inexpensive, quite accurate, and useful for outdoor weighing) (10)

## *Testing Equipment*

soil pH test kits (4, 7, 8)

freshwater dissolved oxygen test kits (10)

freshwater pH test kits (10)

pH test paper (8)

## *Collecting and Examining Equipment*

baby food jars (5, 7, 8, 9, 10)

peanut butter jars (7)

miscellaneous jars and bowls (5, 7, 8, 9, 10)

shallow dishes, droppers, and slides (5, 7, 8, 10)

Baermann funnels (5, 8, 9)

Berlese funnels and lamps (5, 8, 9)

pitfall traps and bait (5)

potatoes (5)

apple corer or knife (5)

meat basters (7, 10)

nutrient agar petri plates (5, 8)

tape recorder and blank tape (9)

white dishes and pans (10)

long-handled collecting net (10)

plankton net (10)

kitchen sieve (small) (10)

aquarium nets (10)

pail (small) (10)

forceps (10)

## *Identification of Organisms*

See Field Guides, pages *xiv–xvii.*

## *Microscopic Equipment*

stereomicroscopes (5, 7, 8, 10)

light microscopes (5, 7, 8, 10)

hand lenses (7, 10)

## *Miscellaneous*

posterboard (or other stiff paper) (1)

felt-tipped markers in assorted colors (1, 2, 3)

tote boxes (4, 5, 6, 7, 8)

trowels (4, 5, 6, 8, 9)

identification tags (3)

wax pencils (3)

marking pens (7)

dowels (¼″) (3, 5, 6, 8)

pruner (to cut dowels) (3)

knife (7)

ball of string (stiff or heavy-duty) (3, 5, 6, 7, 8)

rags or paper towels (5)

wide-mouthed quart jars (mayonnaise) with lids (4, 7)

liquid detergent (4, 5, 7)

mineral oil (5)

masking tape (5, 7, 8, 9, 10)

transparent tape (5)

board (4–6′ long) (1)

cement (small piece) (8)

vinegar (or other dilute acid) (8)

fishing pole (10)

rope and weight (10)

boots or waders (optional) (10)

# General References

## *Outdoor Education*

American Forest Council. *Project Learning Tree. Supplementary Activity Guide, Grades 7–12.* Washington, DC: American Forest Council, 1987.

Garber, Steven. *The Urban Naturalist.* New York: John Wiley and Sons, 1987.

Hammerman, Donald, William Hammerman, and Elizabeth Hammerman. *Teaching in the Outdoors,* 3rd ed. San Francisco: Interprint Press, 1985.

Lawrence, Gale. *Field Guide to the Familiar.* Englewood Cliffs, NJ: Prentice Hall, 1984.

Link, Michael. *Outdoor Education: A Manual for Teaching in Nature's Classroom.* Englewood Cliffs, NJ: Prentice-Hall, 1981.

Mitchell, John, and Massachusetts Audubon Society. *The Curious Naturalist.* Englewood Cliffs, NJ: Prentice-Hall, 1980.

Vogl, Sandra Wolff, and Robert L. Vogl. *Teaching Nature in Cities and Towns.* San Francisco: Interprint Press, 1985.

Western Regional Environmental Education Council. *Project Wild. Secondary Activity Guide.* Boulder, CO: Western Regional Environmental Education Council, 1986.

## *Field Guides (Series)*

*Audubon Society Field Guides.* New York: Alfred A. Knopf.
(detailed but not beyond the beginner, well organized, color photographs for identification)

Bull, John, and John Farrand, Jr. *Audubon Society Field Guide to North American Birds, Eastern Region.* 1977.

Little, Elbert L. *Audubon Society Field Guide to North American Trees, Eastern Region.* 1980.

Little, Elbert L. *Audubon Society Field Guide to North American Trees, Western Region.* 1980.

Milne, Lorus, and Margery Milne. *Audubon Society Field Guide to North American Insects and Spiders.* 1988.

Udvardy, Miklos D.F. *Audubon Society Field Guide to North American Birds, Western Region.* 1977.

*Golden Guide Series.* Racine, WI: Western Publishing Co.
(good basic guides in a handy size, inexpensive, easy to use, and sufficiently detailed for most purposes)

Brockman, C. Frank. *Trees of North America.* 1979.

Fichter, George S. *Insect Pests.* 1987.

Hoffmeister, Donald, and Herbert S. Zim. *Mammals.* 1987.

Levi, Herbert, W., and Lorna R. Levi. *Spiders and Their Kin,* 2nd ed. 1987.

Martin, Alexander C. *Weeds,* 2nd ed. 1987.

Reid, George K. *Pond Life,* 2nd ed. 1987.

Shoemaker, Hurst H., and Herbert S. Zim. *Fishes.* 1987.

Shuttleworth, Floyd S., and Herbert S. Zim. *Mushrooms and Other Nonflowering Plants.* 1987.

Vining, Frank D. *Cacti.* (date not available).

Zim, Herbert S., and Clarence Cottam. *Insects,* 2nd ed. 1987.

Zim, Herbert S., and Ira W. Gabrielson. *Birds.* 1956.

Zim, Herbert S., and Alexander C. Martin. *Flowers.* 1987.

Zim, Herbert S., and Alexander C. Martin. *Trees.* 1987.

Zim, Herbert S., and Hobart Smith. *Reptiles and Amphibians,* 2nd ed. 1987.

*Golden Field Guides.* Racine, WI: Western Publishing Co.
   (larger and more detailed than the above)

Brockman, C. Frank. *Trees of North America.* 1968.

Robbins, Chandler S. *Birds of North America.* 1983.

Venning, Frank D. *Wildflowers of North America.* 1984.

*How to Know Series.* Dubuque, IA: William C. Brown.
   (organized as taxonomic keys, quite detailed, not easy for the beginner)

Chu, H.F. *How to Know the Immature Insects.*

Jahn, Theodore L. *How to Know the Protozoa.*

Prescott, G.W. *How to Know the Freshwater Algae.*

*Peterson Field Guide Series.* Boston: Houghton Mifflin.
   (the standard in the field, but rather intimidating for the beginner)

Borror, Donald J., and Richard E. White. *A Field Guide to Insects of America North of Mexico.* 1970.

Burt, William H., and Richard P. Grossenheider. *A Field Guide to Mammals.* 1976.

Conant, Roger. *A Field Guide to Reptiles and Amphibians of Eastern and Central North America.* 1975.

Craighead, John J., Frank C. Craighead, Jr., and Ray J. Davis. *A Field Guide to Rocky Mountain Wildflowers.* 1974.

Harrison, Hal H. *A Field Guide to Birds' Nests Found East of the Mississippi River.* 1988.

Harrison, Hal H. *A Field Guide to Western Birds' Nests.* 1988.

Krichner, John C., and Gordon Morrison. *A Field Guide to Eastern Forests.* 1988.

Murie, Olaus J. *A Field Guide to Animal Tracks.* 1974.

Niehaus, Theodore F., Charles L. Ripper, and Virginia Savage. *A Field Guide to Southwestern and Texas Wildflowers.* 1984

Peterson, Roger T. *A Field Guide to Birds East of the Rockies.* 1984.

Peterson, Roger T. *A Field Guide to Birds of Texas and Adjacent States.* 1979.

Peterson, Roger T. *A Field Guide to Western Birds,* 3rd ed. 1990.

Peterson, Roger T., and Margaret McKenny. *A Field Guide to Wildflowers of North-eastern and North-Central North America.* 1975.

Petrides, George A. *A Field Guide to Eastern Trees.* 1988.

Stebbins, Robert C. *A Field Guide to Western Reptiles and Amphibians.* 1985.

White, Richard E. *A Field Guide to Beetles.* 1983.

*Peterson First Guide Series.* Boston: Houghton Mifflin.
(abridged and simpflied versions of the above, intended for the beginner)

Alden, Peter, and Richard P. Grossenheider. *Peterson First Guide to Mammals.* 1987.

Filisky, Michael. *Peterson First Guides to Fishes.* 1989.

Leahy, Christopher, and Richard E. White. *Peterson First Guide to Insects.* 1987.

Peterson, Roger T. *Peterson First Guide to Birds.* 1986.

Peterson, Roger T. *Peterson First Guide to Wildflowers.* 1986.

*Stokes Nature Guide Series.* Boston: Little, Brown and Co.
(superb series, emphasizing animal lives more than anatomical detail)

Stokes, Donald W. *A Guide to Bird Behavior,* Vol. I. 1979.

Stokes, Donald W. *A Guide to Nature in Winter.* 1976.

Stokes, Donald W. *A Guide to Observing Insect Lives.* 1983.

Stokes, Donald W., and Lillian Q. Stokes. *A Guide to Animal Tracking and Behavior.* 1986.

Stokes, Donald W., and Lillian Q. Stokes. *A Guide to Bird Behavior,* Vol. II. 1983.

Stokes, Donald W., and Lillian Q. Stokes. *A Guide to Bird Behavior,* Vol. III. 1989.

## *Field Guides (Individual Titles)*

Brown, Lauren. *Grasses: An Identification Guide.* Boston: Houghton Mifflin, 1979.

Dowden, Anne Ophelia. *Wild Green Things in the City: A Book of Weeds.* New York: Thomas Y. Crowell Co., 1972.

Headstrom, Richard. *Suburban Wildflowers.* Englewood Cliffs, NJ: Prentice-Hall, 1972.

Headstrom, Richard. *Suburban Wildlife.* Englewood Cliffs, NJ: Prentice-Hall, 1984.

Hickey, Joseph J. *Guide to Bird Watching*. New York: Dover Publications, 1975.

Peterson, Roger T. *A Field Guide to Bird Songs of Eastern and Central North America*. Cornell University. Records or cassette tapes.

Peterson, Roger T. *A Field Guide to Western Bird Songs*. Cornell University. Records or cassette tapes.

Slavik, Bohumil. *Wildflowers: A Color Guide to Familiar Wildflowers, Ferns and Grasses*. London: Octopus Books, 1973.

Smith, Miranda, and Anna Carr. *Garden Insect, Disease and Weed Identification Guide*. Emmaus, PA: Rodale Press, 1988.

# Guides (Ornamental Plants)

American Horticultural Society. *Illustrated Encyclopedia of Gardening: Fundamentals of Gardening*. Franklin Center, PA: Franklin Library, 1982.

American Horticultural Society. *Illustrated Encyclopedia of Gardening: Shrubs and Hedges*. Franklin Center, PA: Franklin Library, 1982.

American Horticultural Society. *Illustrated Encyclopedia of Gardening: Trees*. Franklin Center, PA: Franklin Library, 1982.

Crockett, James Underwood. *Flowering Shrubs*. New York: Time-Life Books, 1972.

# Natural Areas

Amos, William H. *Limnology*. Chestertown, MD: LaMotte Chemical Products Co., 1969.

*Audubon Society Nature Guides*, Charles Elliott, ed. New York: Alfred A. Knopf.

> Brown, Lauren. *Grasslands*. 1985.
>
> MacMahon, James A. *Deserts*. 1987.
>
> Sutton, Ann, and Myron Sutton. *Eastern Forests*. 1988.
>
> Whitney, Stephen. *Western Forests*. 1985.

Caduto, Michael J. *Pond and Brook: A Guide to Nature Study in Freshwater Environments*. Englewood Cliffs, NJ: Prentice-Hall (date unavailable).

*Our Living World of Nature Series*. New York: McGraw-Hill.

> Allen, Durward L. *The Life of Prairies and Plains*. 1967.
>
> Amos, William H. *The Life of the Pond*. 1967.
>
> McCormick, Jack. *The Life of the Forest*. 1966.
>
> Sutton, Ann, and Myron Sutton. *The Life of the Desert*. 1966.
>
> Usinger, Robert L. *The Life of Rivers and Streams*. 1967.

# Textbooks (College Level)

Alexander, Martin. *Introduction to Soil Microbiology.* New York: John Wiley and Sons, 1977.

Botkin, Daniel B., and Edward A. Keller. *Environmental Studies: The Earth As a Living Planet.* Columbus: OH: Charles E. Merrill Publishing Co., 1982.

Boyd, Robert. *General Microbiology.* St. Louis: Mosby, 1987.

Foth, Henry D. *Fundamentals of Soil Science,* 7th ed. New York: John Wiley and Sons, 1984.

Hickman, Cleveland P., Jr., Larry S. Roberts, and Frances M. Hickman. *Integrated Principles of Zoology.* St. Louis: Mosby, 1987.

Johnson, Leland G. *Biology,* 2nd ed. Dubuque, IA: William C. Brown, 1987.

Miller, G. Tyler, Jr. *Environmental Science: An Introduction,* 2nd ed. Belmont, CA: Wadsworth Publishing Co., 1986.

Northington, David, and J.R. Goodin. *The Botanical World.* St. Louis: Mosby, 1989.

Sherman, Irwin W., and Vilia G. Sherman. *The Invertebrates: Function and Form. A Laboratory Guide,* 2nd ed. New York: Macmillan, 1976.

# Miscellaneous References

Grier, James. *Biology of Animal Behavior,* 2nd ed. St. Louis: Mosby, 1989.

Hancock, Judith M. *Biology Lab Resource Book.* Portland, ME: J. Weston Walch, Publisher, 1985.

Hancock, Judith M. *Project Starters for Biology Classes.* Portland, ME: J. Weston Walch, Publisher, 1988.

McFarland, David, ed. *The Oxford Companion to Animal Behavior.* New York: Oxford University Press, 1981.

# Other Sources of Information

## Professional Publications

*American Biology Teacher*
*Science Teacher*
Supply companies' newsletters and pamphlets

## General Magazines

| | |
|---|---|
| *Audubon* | *Science Digest* |
| *National Geographic* | *Science News* |
| *National Wildlife* | *Scientific American* |
| *Newsweek* | *Time* |

## National Organizations

American Forest Council
1250 Connecticut Avenue NW
Washington, DC 20036

Biology Sciences Curriculum Study
830 North Tejon Street, Suite 405
Colorado Springs, CO 80903

National Audubon Society
950 Third Avenue
New York, NY 10022

National Wildlife Federation
1412 Sixteenth Street NW
Washington, DC 20036

Western Regional Environmental Education Council
Salina Star Route
Boulder, CO 80302

## State/Provincial Organizations

Colleges and universities
    Biology, Botany, Horticulture, Soil Science, Zoology Departments

Environmental organizations (governmental and private)

## Local Organizations

City landscape planners
Environmental organizations
Newspapers
Nurseries
Television news departments

Investigation 1

# Physical Setting
# of the School

## Teacher's Section

This investigation and the following one are introductions to the subject. Their purpose is to acquaint students with the biologically important aspects of the area and to obtain some basic information that will be used in most of the subsequent investigations. The two investigations can be combined into one, if you wish (see Investigation 2).

The school grounds ecosystem, like any other, is made up of both physical and biological entities. The physical ones are the more fundamental, for they form the stage on which the drama of life is played. The physical elements allow life to burgeon with players and actions or restrict their numbers, sometimes even killing the entire cast—usually not for long, however, for life is persistent. For life, the most important physical properties are those that build the stage: light, water, geology. Wherever these necessities are present, life makes its entrance and assumes the lead, modifying the environment as it does so and thus supporting an ever-growing cast. The physical world still calls the tune, however.

In the unique ecosystem of the school grounds, the school building itself is the dominant physical entity on the landscape, influencing the environment in many, often subtle, ways. The most obvious is the pattern of sun and shade which in turn affects soil moisture. Those two factors, light and water, affect all soil organisms, plants included. Because of these pronounced effects, this investigation is focused on the building, including its shape, size, orientation, construction, and the nearby slope of the ground. Some other features, typical of school grounds, are included as well.

The data obtained in this investigation are used to make a map of the area. It is a working model that will be used throughout the study and as such should be large enough to show all the necessary details and to allow more information from subsequent investigations to be added without crowding.

The map should be about three feet square, of stiff paper or posterboard, preferably a single sheet. North or the cardinal compass points (north, south, east, and west) are indicated in an upper corner of the map. The school building, the central feature, is drawn in the middle in its correct shape and orientation. The map need not be to scale, but should accurately reflect the shape of the building and its surrounding features.

Measurements are indicated beside measured structures with double-ended arrows, broken in mid-length for the measurement. The percent of slope (see the following procedure) can be shown on a separate diagram.

Initial drawings should be done in pencil, with colored felt-tipped markers used for the final form. Color-coding can be used for the different types of vegetation (Investigation 2).

Figure 1 illustrates such a map. You may want to make copies for your students as a sample.

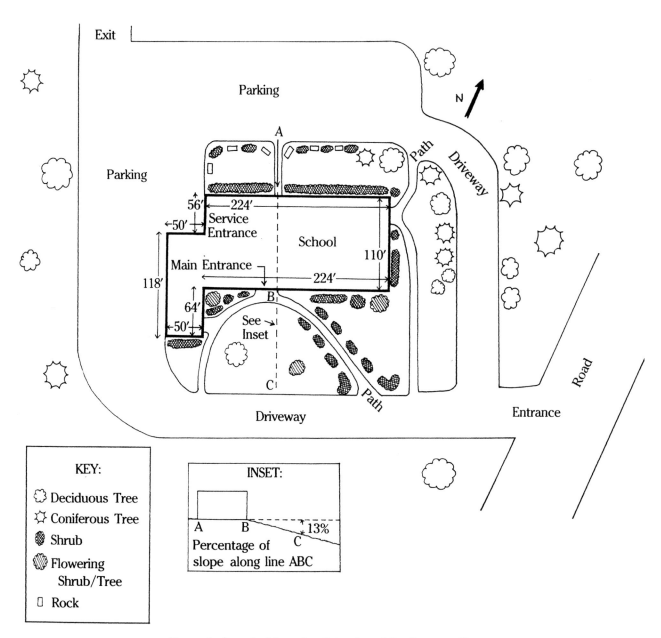

Figure 1. Sample Map of a School and Its Surroundings

(Inset Illustrates Percent of Slope)

# *Procedure*

Divide the class into teams of about three students each and assign each its specific task. Depending on the size of the class and your particular location, it might be a good idea to have more than one team collecting the same data, particularly where measurements are involved.

## 1. Description of the School Building

Students should walk around the building making a sketch of its shape as they go and noting such details as the number of stories, the type of roof (flat or pitched), the amount of overhang, and the location of downspouts, chimneys, doors, stairs, air conditioners, and decks or balconies.

## 2. Measurement of the Building

Measurements are made around the perimeter of the building with accuracy to the nearest inch. To facilitate the process, use a carpenter's long steel tape (20 to 30 feet), if possible. If you don't have one, divide the task among several teams, each with a short (8-foot) tape.

## 3. Orientation of the Building

This measurement, involving the use of the compass, is one that should definitely be done by more than one team; in addition, you may also wish to verify the results by checking on the building's plans in the office.

Compasses come in various levels of precision, with instructions, and the small, inexpensive models sold by supply companies are perfectly adequate for this. You will probably have to show students how to use a compass.

The instrument contains a magnetic needle sealed in a casing with the 16 major bearings, and usually the degrees at each position, inscribed in or on the casing. The magnetic needle points to the magnetic north, not to true north. The two are not the same and the degree of difference depends on your location, but for the purposes of this investigation, magnetic north read directly from the compass is sufficiently accurate. There is also another needle, the directional one, which is used to take a bearing (direction), or the position of an object relative to the magnetic needle.

To use it, the compass is held in the palm of one hand, level and steady so that the magnetic needle can move freely on its own. It will point to magnetic north. South is a straight line away from north, and the east-west line bisects the north-south one at right angles.

First have your students determine north. Then, standing directly in front of the school building and facing it, they move the directional needle so that it points toward the building. From that they can obtain the building's bearing, i.e., its position relative to north, and determine which way it faces. Assume, for example, that the directional arrow is at 250 degrees. That is south of west (west is 270 degrees), so the building faces west-southwest.

## 4. Type of Construction

Students examine the building, noting the kind of materials used in its construction, including foundation, walls, roof, window and door frames, decks, posts, and railings. If you need to, you can verify students' findings with the maintenance department or by checking the building's plans in the office.

## 5. Slope Away From the Building

For many schools, there is no slope. The building is set on a naturally or artificially flat surface, making the slope zero. Where there is some slope, it often involves only part of the building. There is a simple way to determine the amount of slope with ordinary equipment. You need a long board (4 to 6 feet long), a level, and a yardstick. Select a typical slope, put one end of the board on top of it, hold the board steady, and rest the level on it. Raise or lower the free end until the board is level. Hold the yardstick at the free end to measure the distance from the board to the ground and record it to the nearest inch.

The amount of slope is expressed as a percentage: the distance from the board to the ground divided by the length of the board, times 100. For example, if you used a 5-foot board and found the distance from it to the ground to be 8 inches, then the percent of slope is $8/60 \times 100$, or 13%.

In areas of longer slopes, it will be necessary to repeat the measurement, stairstep-fashion down the incline. The forward end of one step becomes the anchor point for the next one down the slope. The slope is then the average of the figures obtained for the several steps.

Slope may also be expressed in degrees for the angle between the horizontal and vertical lengths, using the following formula. It is accurate for slopes of 20 degrees or less, which are the ones most frequently encountered. The figure 57.3 is a constant for those slopes, but not for ones with larger angles, because that figure represents the arc length rather than the vertical length, and it increases as the angle increases.

$$\text{Slope in degrees} = \frac{\text{vertical length}}{\text{horizontal length}} \times 57.3$$

## 6. Location of Driveways, Walkways, Parking Areas, Paths, and Natural Areas

Students make a sketch of the building, showing the location of the areas in relation to it. If your school grounds have distinctive features or natural areas such as a stand of trees or a pond, have students include them as well, since they will be used in a subsequent investigation. Their position relative to the school building is all that is needed at this point.

## *Class Supplies*

| | |
|---|---|
| carpenter's steel tape, 20–30′, or several 8′ tapes | yardstick |
| compass | posterboard or other stiff paper, about 3′ square |
| board, 4–6′ | rulers, pencils, colored felt-tipped markers |
| level | |

## *Spinoff Ideas*

No spinoff investigations are included.

Investigation 1

# Physical Setting of the School

## Student's Section

The area surrounding your school makes up a distinct ecosystem which you'll be investigating for the next several weeks. During that time, you'll come to see the old familiar place in a very different way than you do now, since you'll know it so much better.

This is a very special sort of ecosystem. It is not a natural one such as you would find in a wild area, but an artificial one created around the school. The building itself and everyone who uses it influence all the other living things that share this environment with you. You will be investigating these other inhabitants and the impact of human activities on them, but first you need to learn more about the physical environment.

Your first project is to make a large map of the area. The class will be divided into teams to get the data that are needed, and you'll be using those data throughout your study. Be as accurate as you can.

## *Directions for Obtaining the Data*

### 1. Description of the School Building

Walk around the building, making a rough sketch of its shape as you go and noting such things as the number of stories, the type of roof (flat or pitched), the amount of overhang, and the location of downspouts, chimneys, doors, stairs, air conditioners, and decks or balconies. Then make a final sketch of the building on the data sheet and list all the features you found.

### 2. Measurement of the Building

Measure the perimeter of the building with the steel tape. Make a sketch of the building on the data sheet and record each measurement along the appropriate side on the drawing.

### 3. Orientation of the Building

Your teacher will show you how to use a compass.

When you are familiar with the instrument, stand at the front of the building and determine north. Then stand directly in front of the building, facing it, and move the directional needle on the compass so it points toward the building. This will probably not be the same as the magnetic needle. You can then get the building's bearing, that is, which direction it faces, from the two needles on the compass. You should also be able to determine the building's position in degrees. On the data sheet, sketch the front of the building, show the direction of north by an arrow, and indicate the building's bearing.

### 4. Type of Construction

Examine the building, noting the kinds of materials used in its construction. That includes the foundation, walls, roof, window and door frames, decks, posts, and railings. List this information on the data sheet.

### 5. Slope Away From the Building

Choose several places where there is some slope. Lay the end of the board on the top of the slope and put the level on it. Now raise or lower the free end until the board is level. Place the yardstick at the free end and measure the distance from it to the ground, to the nearest inch. The amount of slope is a percentage: the distance from the board to the ground divided by the length of the board, times 100. For example, if you used a 5-foot board and found the distance to be 8 inches, then $8/60 \times 100 = 13\%$ slope. Record the location and the slope on the data sheet.

If it is a long slope, you'll need to repeat this procedure several times down the incline, so keep track of each place where you set the board. Add up all the percentages you found on this long slope and divide by the number of measurements to give the average slope. Suppose that you made four calculations down a slope at 12%, 13%, 12.5%, and 11%. The average slope is the sum of those numbers (48.5) divided by 4, or 12%.

### 6. Location of Driveways, Walkways, Parking Areas, Paths, and Natural Areas

Walk around the building, making a rough sketch of its shape. Add driveways, walkways, parking areas, and paths. Measurements of paths should also be taken. If there are natural areas at your school, such as a stand of trees or a pond, include them and their position in relation to the school building. Make a final sketch of these features on the data sheet.

## Investigation 1:
# Physical Setting of the School

| *DATA SHEET* |
|---|
| Date: |
| Project: |
| Team Members: |
| Results: |

Investigation 2

# Plant Life on the School Grounds

## Teacher's Section

Biologically, plants are the most important parts of an ecosystem. They form the base of the food web, the producers of both food and oxygen on which all other life depends, and they create the habitats animals require. Because of plants' fundamental role, this investigation focuses on the plants of the school grounds.

The school grounds' ecosystem is a distinctly artificial one. When the building was built, the land was leveled or gently sloped; service areas such as driveways, walkways, and parking areas were placed in convenient proximity to the building; and landscaping around the building and service areas was added. The purpose of the landscaping is aesthetic.

Typical landscaping involves foundation planting, borders along walkways, and free-standing trees and shrubs, with a carpet of lawn uniting these areas and often extending well beyond them. There are, of course, many variations depending on locale; urban schools have limited grounds, while suburban ones often have extensive, well-landscaped grounds.

Usually the plant species selected are horticultural types that have been developed specifically for landscaping. They are termed **cultivars**, a contraction of "cultural varieties," which were created from wild species by horticulturalists and are characterized by being hardy, easily cared for, attractive, and not too expensive. Native species, particularly trees, are often used as well. In terms of identification, though, cultivars and native species are not found in the same references; field guides are used for wild species and gardening or horticulture books for cultivars.

Foundation plants are those planted around the foundations of a building to mask the foundations and tie the building visually to the land. The most common types of foundation plants are the evergreen shrubs—in particular, needled species (juniper, yew, dwarf arborvitae) and broad-leafed flowering types (azalea, rhododendron). Deciduous shrubs (barberry, cotoneaster, viburnum) are also used, and vines (ivy, wisteria) are found, especially on old buildings. Walkways are often bordered with a continuation of the foundation planting.

Free-standing trees and shrubs are individual specimens considered as accent pieces on a lawn. Some of the more common ones are the flowering shrubs (forsythia, lilac, viburnum); small flowering trees (crabapple, dogwood, hawthorn); large deciduous shade trees (maple, oak, white birch, beech, willow, elm); and evergreen shade trees (spruce, pine). Some genera have species with different growth habits, such as cherry (trees, shrubs), euonymus (trees, shrubs, vines), and honeysuckle (shrubs, vines); all are commonly used landscape plants.

The landscaped area immediately surrounding the school building is the subject of this investigation. Natural or seminatural areas found on many school grounds will be considered in Investigation 10.

Here, students will be identifying the various kinds of landscape plants. Cultivars can be difficult to identify, and identification need not be scientifically exact; common generic names are perfectly adequate. Should you run into problems with some plants, cut off a small twig and take it to a nursery or gardening center for help.

Students will be working in different areas. Their results should be correlated into a single list of all the plants, and their locations are shown on the map. Symbols and color-coding are the easiest way to represent the different types of plants on the map (see Figure 1). Both the list and the map are references for subsequent investigations.

## *Procedure*

Depending on how extensive the plant life is at your school grounds, you might be able to combine this investigation with the preceding one, since both are introductory and are handled in much the same way. Divide the class into teams of about three students each and assign each its specific task. If you have a large building and/or a great many plants, you might wish to subdivide the areas further.

### 1. Foundation Planting

First have your students measure the distance from the foundation to the base of each plant as well as between plants. This should be done to the nearest inch; yardsticks or 8-foot steel tapes are adequate. The plants are then identified.

### 2. Walkway Borders

Follow the same procedure as above.

### 3. Free-Standing Trees and Shrubs

Students measure the shortest distance from each plant to the building, to the nearest inch, and then identify the plants. Carpenter's long steel tapes are handy for this one.

### 4. Flower Beds

Few schools have flower borders in front of foundation planting or along walkways, but some do have planters. If yours does, students can identify the plants, typically annuals.

### 5. Lawn

The amount of lawn varies greatly with locale. If yours has lots of lawn, limit the project to the part near the building and the adjacent service areas. Lawns usually do not consist of grass alone but contain assorted weeds as well (Investigation 6). However, for the present, students should treat the lawn as a single entity and measure its extent in the areas in question. Often paths cut across the lawn in several places, and they too can be measured.

## *Class Supplies*

guides (see general reference list, pages *xii–xv*)
yardsticks, 8′ steel tapes, or 20–30′ tapes
map (See Investigation 1)
rulers, pencils, colored felt-tipped markers

## *Spinoff Ideas*

No spinoff investigations are included.

## *References*

American Horticultural Society. *Illustrated Encyclopedia of Gardening: Fundamentals of Gardening.* Franklin Center, PA: Franklin Library, 1982.

American Horticultural Society. *Illustrated Encyclopedia of Gardening: Shrubs and Trees.* Franklin Center, PA: Franklin Library, 1982.

American Horticultural Society. *Illustrated Encyclopedia of Gardening: Trees.* Franklin Center, PA: Franklin Library, 1982.

Bienz, D.R. *The How and Why of Home Horticulture.* San Francisco: W.H. Freeman Co., 1980.

Crockett, James Underwood. *Flowering Shrubs.* New York: Time-Life Books, Inc., 1972.

Garber, Steven. *The Urban Naturalist.* New York: John Wiley and Sons, 1987.

Headstrom, Richard. *Suburban Wildflowers.* Englewood Cliffs, NJ: Prentice-Hall, 1972.

Nadel, Ira Bruce, and Cornelia Hahn Oberlander. *Trees in the City.* New York: Pergamon Press, 1977.

Investigation 2

# Plant Life on the School Grounds

## Student's Section

In the first investigation, you determined some important physical aspects about the ecosystem on the school grounds. In this one you are going to continue your general description of the environment by examining the plant life. Plants are the most important, and most obvious, parts of the biological environment. Again, you will be working in teams. Your results, along with those of the rest of the class, will be added to the map.

### *Directions for Obtaining the Data*

#### 1. Foundation Planting

Measure the distance from the foundation to the base of each plant and the distance between plants, to the nearest inch. On the data sheet make a sketch of the base of the building showing where the foundation plants are. Then identify the plants and list them on the data sheet.

#### 2. Walkway Borders

Follow the same procedure given for foundation planting.

#### 3. Free-Standing Trees and Shrubs

Measure the shortest distance from each plant to the building, to the nearest inch. On the data sheet make a sketch of the base of the building and show where these plants are in relation to it. Identify the plants and list them on the data sheet.

#### 4. Flower Beds

Identify the plants in flower borders or planters and list them on the data sheet. Include sketches to show their location in relation to the building.

#### 5. Lawn

Measure the extent of lawn in the area assigned to you. If paths cut across the lawn, measure them as well. On the data sheet, make a sketch of the building with its service areas, showing the lawn around them.

# Investigation 2:
# Plant Life on the School

| DATA SHEET |
| --- |
| Date: |
| Project: |
| Team Members: |
| Results: |

# Investigation 3

# Health of the School Grounds' Plants

## Teacher's Section

Often the landscaping around a school building is done at the time the building is built, and aside from mowing the lawn and clipping hedges, scant attention is paid to it thereafter. The species selected, generally, are those able to withstand the rigors of a school environment, though that is not always true, and most do require more care than they receive. As a result, the plants are in various stages of health, or lack of it. Plant health, a good subject for investigation, is not a topic considered in biology courses.

Healthy plants are green. The shade of green varies considerably, with young leaves, particularly in the spring, being a paler green than the mature leaves on the same plant. There are species differences, too, ranging from medium to dark. Nevertheless, the color of a healthy plant is a definite, unmistakable green. The stems of herbaceous plants are green, whereas those of of woody ones are grays and browns (except the white birch). The stems of herbaceous plants and shrubs and the small branches of trees are pliable, bending and tearing under pressure rather than breaking. A dead stem or branch of the same size snaps off readily. Healthy tree trunks are intact, without holes or parasitic growths, such as lichens or fungi, on them. As healthy plants grow, they develop shoot and flower buds, usually at the stem tips. Shoot buds, the growth centers for woody stems, contain the meristematic tissues responsible for stem elongation and leaf formation. Flower buds, too, follow a normal course of development, from opening to maturation of pistil and stamens, pollination, and finally seed and fruit formation.

The life of a plant has three stages: juvenile, mature, and senescent. The seedling, dependent on the food reserves in the seed and/or cotyledons, soon begins photosynthesis and that point, when the young plant is on its own, is the beginning of the juvenile stage. It lasts until reproduction begins. It is a period of high metabolism (photosynthesis and respiration) and rapid synthesis of organic compounds and organelles, of cell division and growth. In some species, juvenile leaves may be quite different in appearance from mature ones.

Plants in the mature stage are actively reproducing, usually at a characteristic time during the year, and it is then that they develop reproductive structures (flowers in angiosperms, cones in gymnosperms). Mature plants continue to have a high metabolism, but the growth rate is slower than in juvenile plants because many of the organic compounds, especially carbohydrates and lipids, are stored in reproductive structures rather than being used for growth.

*13*

Metabolic changes in the older mature plant signal the beginning of senescence, the period in which there is a decline in metabolism and the synthesis of organic compounds and organelles. Senescence in annuals and biennials normally begins after the seeds have matured and involves the gradual deterioration of the whole plant. Interestingly, it can be delayed by preventing seed maturation (plucking off dead flowers or young seeds). The life cycle in perennials is much longer, and their responses are also different. In the fall in cold climates, the leaves and shoots of herbaceous perennials (grass, for example) die, but the roots and crown (stem-root junction) remain living and rejuvenate in the spring. Senescence, thus, does not necessarily lead to death. Many woody perennials, such as shrubs, do not appear to age as they annually put forth new shoots from the roots or crown. Trees, however, do undergo senescence. As the metabolic processes slow down, individual branches may die, and various parasitic and saprophytic organisms begin to live on, or in, the tree. These organisms, mainly insect larvae, lichens, and fungi, draw their sustenance from the tree. This weakens the tree and tends to speed up the inherent senescent processes. Weather, too, takes its toll. Finally, the tree dies.

There are many types of plant diseases. Diseases are caused by organisms, nutrient deficiencies, and injuries. Diagnosis usually is difficult, since many diseases have similar symptoms. Accurate diagnosis is the province of the professional plant pathologist, but students can learn to distinguish healthy plants from unhealthy ones and to recognize disease symptoms. These are the most common symptoms:

1. Extensive growth of lichens, fungi, and mosses on the trunks and branches of trees. ("Extensive growth" means that it covers most of the bark, as opposed to scattered patches of growth.)
2. Open holes in trees used as dens by small animals, which enlarge the holes in the process.
3. Stunted or spindly growth—in the case of grass, the failure to spread out normally, resulting in bare patches or sparse covering of the soil.
4. Yellowing of leaves and herbaceous stems (chlorosis).
5. Browning and loss of leaves or needles at the wrong season of the year.
6. Poor reproduction (flowers, seeds, or fruits fail to form or drop prematurely).
7. Tissue or organ death (necrotic spots in leaves, shoot tips, flowers, fruits).
8. Galls and tumors (galls are growths of organized tissues on leaves or stems, while tumors are unorganized growths).

The principal disease-causing organisms in plants are viruses, bacteria, and fungi, of which the fungi are the largest and most important group. Pathogens act both intracellularly (viruses and bacteria), by interfering with the metabolic machinery of the cell, generally by affecting enzyme production and function, and extracellularly (fungi), mainly by blocking the vessels of the transport system. The responses of the plant to the presence of these pathogenic organisms, or their toxic products, may be overt, such as the symptoms on the preceding list, or may be subtle and discernible only biochemically.

Disease is a fact of life, one of interest to many people and well worth studying. The subject is not pursued in school because of the fear of diseases being transmitted to students. However, disease-causing organisms are extremely host-specific; plant pathogens cannot affect people, and so diseased plants are completely safe to handle.

Mineral-deficiency diseases are caused by the lack of one or more of the vital minerals. The minerals may be absent from the soil or present but unavailable to the plant due to the acidity or alkalinity of soil water (see Investigation 4). Symptoms of mineral deficiency include:

- chlorosis (yellowing) due to deficient chlorophyll synthesis
- necrotic spots (white, brown, or black) on leaves
- spindly or stunted growth
- curling, wilting, or premature loss of leaves
- inability to produce flowers or fruit
- deformities due to defective meristematic activities

Injuries to plants are caused by the activities of animals; insects are the major culprits. Insect injury is caused by:

- eating (chewing or piercing and sucking fluids)
- egg-laying (larvae then eat the plant from within)
- acting as vectors for microbial disease organisms (deposited in wounds made by eating or egg-laying).

Seven orders of the class Insecta contain species that cause injury to plants:

- Orthoptera (grasshoppers, crickets)
- Thysanoptera (thrips)
- Hemiptera (bugs)
- Coleoptera (beetles)
- Lepidoptera (moths, butterflies, particularly as caterpillars)
- Diptera (flies)
- Hymenoptera (bees, ants, wasps, sawflies)

Another major group of animals that injure plants are the nematode worms (Investigation 5). Most of them live in soil and feed on roots, causing tissue death, reduction of the root system, rotting, and galls in some cases. Other injury-causing animals are den-builders (mice and squirrels), birds (particularly woodpeckers), and urinating dogs.

Human activities, ranging from absentmindedly breaking off twigs to soil compaction, construction work, vehicles, and pollutants, also affect plants (see Investigation 9).

# *Procedure*

This investigation calls for closer examination of the plants that were studied in Investigation 2. It includes the trees and shrubs of the foundation planting, walkway borders, and free-standing specimens, as well as the grasses of the lawn near the building.

This is a good sharing activity, so plan on students working together in pairs. You can use two approaches to assigning sites: either divide the area randomly or select individual plants. Both approaches have advantages and disadvantages. Which one you choose depends on the extent of planting, the size of your class, the amount of time, and your own approach to the subject.

If you divide up the area randomly, each pair of students will have a certain number of plants to examine. This is less time-consuming for you, but the outcome is less certain. All the plants may be healthy, or not, but there is apt to be considerable difference among plants and therefore in student workloads. On the other hand, if you wish to select individual plants, you have to check them ahead of time, making sure you have good examples and that the workload is equalized as far as possible. This, obviously, takes more of your time. The major drawback is that you can easily distort the outcome if you concentrate too heavily on unhealthy plants.

The only way students can diagnose how healthy plants are is by comparison. Ideally, that is done with plants of the same species and age, an unlikely situation at a school, but working with the available plants, students can still make their comparisons and diagnoses.

The number of plants to be examined by each pair of students depends both on the number of plants and the amount of time you wish to spend on this investigation, but three or four trees and shrubs (either or both) and two grass plots are sufficient. One data sheet is needed for each tree or shrub and one for two grass plots. You do not have to divide the sites equally; generally, more partnerships are needed for trees and shrubs than for grass.

Students need to distinguish their sites and specimens. The easiest way to mark off sites is to use four stakes as corner markers and a piece of string to measure distance. Stakes can be 6 to 8 inches long, of ¼″ dowel, which can be cut with a pruner and sharpened in a pencil sharpener to make them easier to stick in the ground. Individual specimens are tagged with the name of the species and an identification number and the tags tied to a branch. Use tags with metal rings and a wax pencil so the tags last for the duration of the study. The easiest way to set up a numbering system without overlaps is for you to assign the numbers, giving each pair of students more numbers than you expect them to need (don't worry if they aren't all used).

## 1. Trees and Shrubs

The procedure is the same for these plants wherever they are located. Students should first tag each specimen with the species name (consult the list made in Investigation 2 if need be) and a number, so it can be identified again. *Pliability*, a term used on the data sheet, refers to bendability. Holding a twig in both hands, you gently bend it backwards. One that is pliable can be bent and will snap back when the pressure is released. Dead or dying twigs break.

In examining trees or tall shrubs, have students first stand back to see the overall shape, a species characteristic, though it is often modifed by pruning. Then students examine each specimen closely, following the list on the data sheet, and recording their observations as they do so. The data sheet also calls for students to draw conclusions and to make diagnostic judgments based on their observations.

## 2. Grass

Students should first notice the appearance of the whole lawn, then select two contrasting areas, each about three feet square, but not including paths. The areas are then staked out so they can be identified again. Students then examine the grass in the two areas carefully, following the list on the data sheet, recording their observations as they do so and drawing conclusions.

*Thickness* and *thatch* are two terms used on the data sheet with reference to lawn. Thickness refers to the number of shoots the plants put out from their crowns. A healthy, thick lawn has so many shoots that it is hard to see the soil, but a sparse lawn has fewer shoots and the soil shows through between the plants. Thatch is the collection of clippings and dead grass on the surface of the soil. A lot of thatch blankets the surface, cuts off air to the roots, may retain moisture, provides a good habitat for insects, and as a result, tends to promote disease.

# Results

Individual students only see a small part of the whole. In order to gain some perspective, they need to find out what others discovered and think. To facilitate comparisons, organize this reporting by the type of plant (trees, shrubs, grass). Specimens of particular interest can be added to the map using a special color code.

Of necessity, this session will be quite factual. Students may well raise questions of a causative nature—indeed, they should—but at this point there are no answers. Several subsequent investigations, especially Investigations 4 and 8, will provide information that will help answer students' questions—and raise a few more.

It doesn't hurt if students' questions go unanswered. You can let the students speculate, if you wish, or advise them to "stayed tuned." Everything cannot be done at once; anticipation and tying things together are educationally worthwhile.

## Supplies for Each Partnership

| | |
|---|---|
| 4–5 tags (with metal rings) | string, 3′ long |
| wax pencil | plant list from Investigation 2 |
| 4 stakes, 6–8″, sharpened (¼″ dowel) | |

## Class Supplies

| | |
|---|---|
| map (from Investigations 1 and 2) | colored felt-tipped markers |

## Spinoff Ideas

No spinoff investigations are included.

## References

American Horticultural Society. *Illustrated Encyclopedia of Gardening: Fundamentals of Gardening.* Franklin Center, PA: Franklin Library, 1982.

Brooks, Audrey, and Andrew Halstead. *Garden Pests and Diseases.* New York: Simon and Schuster, 1980.

Pirone, Pascal P. *Diseases and Pests of Ornamental Plants,* 4th ed. New York: Ronald Press Co., 1970.

Smith, Miranda, and Anna Carr. *Garden Insect, Disease and Weed Identification Guide.* Emmaus, PA; Rodale Press, 1988.

Investigation 3

# Health of the
# School Grounds' Plants

## Student's Section

Your teacher has already discussed with you the appearance of healthy plants and the symptoms of diseased ones, so you know the difference. Now you're ready to apply that information to some plants and diagnose how healthy they are. In order to do that, you'll have to examine the plants carefully and draw some conclusions about them.

For reference, these are the symptoms of the most common diseases in plants:

1. Extensive growth of lichens, fungi, and mosses on the trunks and branches of trees. ("Extensive growth" means that it covers most of the bark, as opposed to scattered patches of growth.)

2. Open holes in trees used as dens by small animals, which enlarge the holes in the process.

3. Stunted or spindly growth—in the case of grass, failure to spread out normally, resulting in bare patches or sparse covering of the soil.

4. Yellowing of leaves and herbaceous stems (chlorosis).

5. Browning and loss of leaves or needles at the wrong season of the year.

6. Poor reproduction (flowers, seeds, or fruits fail to form or drop prematurely).

7. Tissue or organ death (dead spots in leaves, shoot tips, flowers, fruits).

8. Galls and tumors (galls are growths of organized tissues on leaves or stems, while tumors are unorganized growths).

         *Biology Is Outdoors!*

# Directions for Obtaining the Data

## 1. Trees and Shrubs

Tag each specimen at your site with the species name and a number, so you can identify it again. If you have trees or tall shrubs, first stand back where you can see the overall shape. That gives you an idea of what it looks like as a whole even though its shape may have been modified by pruning. Then examine each specimen closely, following the list on the data sheet. The term *pliability* (on the data sheet) refers to bendability. Take hold of a twig in both hands and gently bend it backwards. One that is pliable can be bent and will snap back when pressure is released. Dead or dying twigs break.

Record your observations on the data sheet at the time you make them and then draw your conclusions. Each plant requires a separate sheet.

## 2. Grass

You can hardly look at each individual grass plant, so you'll have to select areas to study. First notice the appearance of the whole lawn, then select two areas, each about three feet square, and stake them out. Be sure the grass in one area looks different from the grass in the other area. Don't include paths. Examine the grass in the two areas carefully, following the list on the data sheet.

You need to know two terms: *thickness* and *thatch*. Thickness refers to the number of shoots the plants put out from the crown (stem-root junction). A healthy, thick lawn has so many shoots that it is hard to see the soil, but a sparse lawn has fewer shoots and the soil shows through between the plants. Thatch is the collection of clippings and dead grass on the surface of the soil. A lot of thatch blankets the surface, cuts off air to the roots, may retain moisture, provides a good habitat for insects, and as a result, tends to promote disease.

Record your observations on the data sheet at the time you make them and then draw your conclusions. You need one data sheet for two plots.

Name _____  Date _____

# Investigation 3:
# Health of the School Grounds' Plants

| **DATA SHEET**—Trees and Shrubs |
|---|
| Date: |
| Site: |
| Partner: |
| Species:                                    No.: |
| Leaves/Needles: |
| *Color:* |
| *Loss:* |
| *Insects:* |
| *Evidence of Being Eaten:* |
| *Overall Condition:* |
| Twigs: |
| *Tips/Buds:* |
| *Evidence of Growth:* |
| *Pliability:* |
| *Overall Condition:* |
| Branches: |
| *Most Living:* |
| *Many Dead:* |
| *Overall Condition:* |
| Stems/Trunks: |
| *Color:* |
| *Wounds:* |
| *Lichens/Fungi/Mosses:* |
| *Overall Condition:* |
| Overall Health: |
| Good              Fair              Poor              Dying |
| (Circle the term that most accurately describes the plant's overall health, or put a check mark on the line where it belongs between two terms.) |

Name _____ Date _____

# Investigation 3:
# Health of the School Grounds' Plants

| *DATA SHEET*—Grass |
|---|
| Date: |
| Site: |
| Partner: |
| Area 1: _____ |
|    *Color:* _____ |
|    *Thickness:* _____ |
|    *Thatch:* _____ |
|    *Insects:* _____ |
|    *Evidence of Being Eaten:* _____ |
|    *Non-Grass Species:* _____ |
|      Fungi: _____ |
|      Moss: _____ |
|      Clover: _____ |
|      Weeds: _____ |
| Overall Health: |
|     Good            Fair          Poor        Dying |
| Area 2: _____ |
|    *Color:* _____ |
|    *Thickness:* _____ |
|    *Thatch:* _____ |
|    *Insects:* _____ |
|    *Evidence of Being Eaten:* _____ |
|    *Non-Grass Species:* _____ |
|      Fungi: _____ |
|      Mosses: _____ |
|      Clover: _____ |
|      Weeds: _____ |
| Overall Health: |
|     Good            Fair          Poor        Dying |
| (Circle the term that most accurately describes the grass's health, or put a check mark on the line where it belongs between two terms.) |

Investigation 4
# Soil Analysis

## Teacher's Section

Soil is an enormously complex ecosystem composed of weathered rock particles, organic materials in various stages of decay, soil organisms, air, and water. Organic materials are broken down to inorganic ones by the organisms and in combination with the rock particles comprise the substances (ions, radicals, and molecules) that collectively are known as soil minerals.

Rocky particles vary in size from clay (particle size less than 0.002 mm diameter) to silt (particle size 0.002 to 0.05 mm diameter) to sand (particle size 0.05 to 2.0 mm). The spaces surrounding the particles are filled with air or water. The size of the particles and their angularity determine the amount of space around them, since tinier particles pack more closely than larger and more irregular ones do. The amount of space determines air- and water-holding capacities. Clay absorbs water readily, with particles and spaces coated, and becomes waterlogged. That reduces or even eliminates the amount of air in clay soil. Sand, on the other hand, has large spaces, holding plenty of air, but since water tends to percolate through the spaces easily, sandy soil has poor water-retaining properties. Neither sand nor clay makes good soil. The best growing soil, loam, contains clay and sand particles in roughly equal amounts, along with about the same amount of organic material, or humus. Humus gives the soil its bulk. It is nutrient-rich, alive with organisms, and spongy, retaining minerals, air, and water in its loose structure.

In order to sustain their lives normally, plants require six macrominerals (nitrogen, phosphorus, potassium, calcium, magnesium, and sulfur) and seven microminerals (iron, copper, zinc, molybdenum, cobalt, boron, and manganese). They make up those essential compounds, DNA, RNA, and proteins (both cell structures and enzymes), as well as many other vital compounds such as chlorophyll.

Minerals enter the roots not as solids but as ions dissolved in soil water. Their solubility depends on the pH of the soil water, which in turn is influenced by the chemistry of the specific soil. Each mineral has its own solubility range: an optimal pH range at which it is most soluble, gradually tapering off both above and below that level until it is no longer soluble. All the macrominerals, for example, are insoluble at pH 4 and below.

Actually, the situation is more complicated. Many more minerals are in the soil than plants require, and under the proper conditions these minerals react to form more complex, and usually insoluble, compounds. Molybdenum and iron, for example, react at low pH levels to form an insoluble compound, making both of them unavailable to plants.

The total range of pH values for plant life is 4 to 9, but the species that can live at the extremes are few. They are found, respectively, in peat bogs and alkali deserts, neither of which concern us here. Most garden soils are in the range of 6 to 8 pH, a range that coincides with optimal solubility of all the vital minerals. Not surprisingly, plants do well under those conditions.

Many North American soils, however, fall in the pH range of 5 to 6. These acid soils are usually treated with lime to raise the pH. As another example of the complexity of soil interactions, many minerals, required or not, are quite soluble at low pH levels. Excesses are absorbed and can easily reach a toxic level in the plant body (another reason why most plants don't thrive when the pH is low). As the pH is raised toward neutrality, the solubility of these minerals decreases, thus eliminating the excess and preventing toxicity. If, however, the pH continues to be raised in order to make some minerals more soluble, then the solubility of other minerals may decrease to the point of deficiency.

Aluminum sulfate is used to lower the pH. As is true with lime, altering the pH affects mineral solubility, sometimes to the point of creating problems.

Different species of plants have different mineral requirements and therefore require different types of soil, but there are some common aspects of plant physiology. Plants absorb all the ions, radicals, and molecules present in soil water whether they are needed or not (size is the only limit). Since plants cannot distinguish beneficial from harmful substances and cannot exclude soil water substances below a certain size, mineral concentrations in plants can reach toxic levels. Minerals toxic to plants include: aluminum (the most common mineral), arsenic, heavy metals (lead and mercury, particularly), and essential microminerals in high concentrations. Toxicity results from mineral imbalance.

# *Procedure*

Divide your class into teams of about three members each and assign each a site, using the same sites as in Investigation 2 (foundation planting beds, walkway borders, near free-standing trees or shrubs, and lawn) and adding paths that cut across the lawn. In order to examine the soils so that comparisons can be made among different sites, each team should do all the tests. The results are then recorded on the data sheets and shared with the rest of the class (see "Results").

The tests are done on soil below the surface. This necessitates digging down several inches, into the root zone. Digging on the lawn means removing a clump of sod about 6 inches in diameter. It should be placed in the shade, roots down, while the investigation is going on and then replaced, tamping it down carefully.

## 1. Preliminary Examination

Students should examine the soil by looking closely at a handful of it. The different components are quite apparent. Minerals are light-colored, rocky, and angular. Humus is dark and contains fragments of twigs or other recognizable bits of biological material.

### a. COMPOSITION

Students determine, subjectively, the relative amounts of minerals and humus present.

### b. FRIABILITY

This is an informal gardener's test to determine when soil is in good planting condition. Pick up a handful of soil and squeeze it tightly. When the pressure is released, good friable soil slowly crumbles apart, but wet soil holds together in a ball and dry or sandy soil simply runs through your fingers.

## 2. Mineral Content

The relative proportion of the different minerals can be determined quite easily. If soil is mixed with water and then allowed to stand, the particles will settle out in layers according to weight. Sand particles are the heaviest and settle out quickly, while lighter-weight silt and clay particles take much longer. They gradually settle to form two layers above the sand, with silt in the middle and clay on top. However, the settling of these fine particles, particularly clay (which is held in suspension by the motion of water molecules), is too slow for classroom purposes. A wetting agent, such as is present in liquid detergents or water softeners, speeds up the process by coating the particles and thus altering their natural behavior.

Provide each team with a wide-mouthed quart jar (mayonnaise ones are good). Students add water until the jars are slightly over half full, then put in one to two teaspoonfuls of detergent, and then fill the jar almost full of soil. Next, students screw the cap on tightly, shake the jar vigorously to mix the contents thoroughly, and then let the jar stand.

Settling begins almost at once, and the layers gradually appear as the water clears.

Separation involves humus particles as well as the inorganic minerals. The former are scattered in the different layers according to their weight, and probably some of the lightest will continue to float even after the minerals have settled. Mineral separation makes it possible to identify the different minerals by their behavior and appearance. If you wish to examine them more closely, you can insert a pipette into the layer, bulb closed, then open it to extract a sample. The samples can be examined microscopically.

The jars, labeled with the site they came from, can be kept for reference throughout the study.

## 3. Water-Holding Capacity

One of the most important biological properties of soil is its capacity to retain water. There are several ways to determine this, but usually what is measured is the opposite: drainage. All soils retain water in the spaces around particles, and in all soils excess water drains downward, though the speed with which it does so varies greatly. Humus, which has a large surface area, retains water well, and the more of it there is, the better the water-holding capacity. Little drains off. Sandy soils, at the opposite extreme, are so porous that water practically flows through them unimpeded. Thus the easiest way to determine water-holding capacity is to measure the speed of drainage (that is the "perk test" done on housing sites).

Traditionally, the test consists of pushing a coffee can, both ends removed, into the soil, filling it with water, and timing how long it takes the water to drain out—a procedure fraught with obvious hazards.

Instead, use a PVC coupler or a length of PVC tubing about four inches in diameter. The coupler, a smooth, finished piece, is about four inches long. Tubing can be cut to any length (about six inches is good), but after cutting you do have to make sure the ends are smooth. It doesn't matter which piece of PVC you use, nor the exact size, but the teams should all have identical material so that results can be compared.

Students stand the tube on end on the ground and push it about two inches down into the soil. That is easily done in cultivated soil, but may be next to impossible on lawn or packed paths. In those cases, have students cut a circle slightly larger than the outside of the tube with the trowel and set the tube in the cut. It should be firmly in place. If not, students should pack some soil around the outside but leave the soil inside undisturbed. Students fill their tube with water and record the time when it was poured in and when it had all drained out. That amount of time is a measure of the soil's water-holding capacity.

## 4. pH

For you, the easiest thing to do is to use soil pH kits purchased from a biological or garden supply company. They are quite inexpensive and have easy-to-follow directions. Ideally, provide each team with a kit, but if you can't afford that, two teams can share a kit, or you can subdivide one.

Many kits provide tests for the major macrominerals (N, P, K) as well as pH. While you might wish to have several of these kits available, it is not necessary for simple pH tests.

Emphasize to your students that pH is a logarithmic scale. Each whole number differs from the next by a factor of 10 (i.e., pH 5 is 10 times more acidic than pH 6 and 100 times more acidic than pH 7).

## *Results*

As was true in the last investigation, individual students have seen only part of the picture and need to share their findings in order to complete the picture. Comparisons among the different sites are crucial to student understanding. Comparisons can be made in one of two ways.

You can have each team in turn report its results, by putting the data sheet information on the board. Then the class can compare the different sites and draw conclusions. Or you might compile a composite table from all the data sheets, which you can duplicate and hand out prior to discussing the results. Whichever method you choose, you should preserve the results in a composite table, following the style of the data sheet. The composite will be used in subsequent investigations. Post a copy beside the map. You might want to include some of the data from this investigation on the map.

## *Supplies for Each Team*

| | |
|---|---|
| tote box | PVC coupling or tubing, 4″ diameter |
| trowel | watch |
| quart jar (wide-mouthed) with cap | soil pH test kit |
| liquid detergent and teaspoon | |

## *Class Supplies*

| | |
|---|---|
| bucket of water and dipper or hose | map (from previous investigations) |

## *Spinoff Idea*

Suggestion for a spinoff investigation:

• Effects of Leaching and Erosion on Plant Growth

# *References*

American Horticultural Society. *Illustrated Encyclopedia of Gardening: Fundamentals of Gardening.* Franklin Center, PA: Franklin Library, 1982.

Bienz, D.R. *The How and Why of Home Horticulture.* San Francisco: W.H. Freeman Co., 1980.

Foth, Henry D. *Fundamentals of Soil Science,* 7th ed. New York: John Wiley and Sons, 1984.

Foth, Henry D. *Study of Soil Science.* Chestertown, MD: LaMotte Chemical Products Co. (date unknown).

Investigation 4

# Soil Analysis

## Student's Section

This investigation is concerned with some of the physical properties of soil that are important to organisms. Different teams will be assigned to different sites so that the variations in the soils on the school grounds can be determined and then compared.

Select one spot on your site to study the soil. All the tests are done on soil below the surface, so dig a hole several inches down. Pile the soil to one side. If you dig on the lawn, remove a clump of sod about six inches in diameter and set it aside, in the shade, with the roots down. When you've finished, replace the soil and sod in the hole and tamp it down carefully.

Record the results of each test on the data sheet as soon as you have them.

## *Directions for Obtaining the Data*

### 1. Preliminary Examination

Pick up a handful of soil and examine it closely. You should be able to see the different components. Minerals are light-colored, rocky, and angular. Humus is dark and contains fragments of twigs or other recognizable bits of biological material.

#### *a. COMPOSITION*
Determine the relative amounts of minerals and humus present.

#### *b. FRIABILITY*
*Friability* refers to the texture of the soil, **not** whether it can be fried. It's a gardener's term. Pick up a handful of soil, squeeze it tightly, then open your fingers. If the soil is powdery (dry or sandy with low organic content), it will run through your fingers. If the soil forms a ball that crumbles when you release the pressure, it has a high organic content and is said to be "friable." If the ball holds together, it is wet or has a high clay content.

## 2. Mineral Content

This test is based on the rate of settling of different minerals from a water-soil mixture.

Pour water into the quart jar until it is about half to two-thirds full. Add one to two teaspoonfuls of detergent to speed up the settling of the tiniest particles. Swirl to mix the detergent with the water, then add soil until the jar is almost full. Screw the cap on tightly and shake the jar vigorously so the contents are thoroughly mixed. Let the jar stand while you do the other tests. Then examine the layers that formed in the jar and determine which minerals are in which layers. Do that without stirring them up.

## 3. Water-Holding Capacity

While the lives of all soil organisms depend on the ability of the soil to hold water, that is difficult to measure. It is easier to determine drainage. The two properties, water-holding and drainage, are opposites, found in all soils but differing in rate among different soil types.

Stand the PVC tube on end and push it about two inches down into the soil. That's easy on cultivated soil and next to impossible on lawn or paths. For them, mark a circle with the tube, cut down with the trowel to form a channel, and set the tube in it. The tube should be about two inches down into the soil and firmly in place. If it isn't firmly set, pack soil around the outside of the tube, but don't disturb the soil on the inside. Fill the tube with water and record the time when the water was poured in and when it had all drained out. That amount of time is a measure of the soil's water-holding capacity.

## 4. pH

Follow the pH test kit's directions.

Name _____  Date _____

## Investigation 4:
# Soil Analysis

| *DATA SHEET* |
|---|
| Date: |
| Site: |
| Team Members: |
| 1. Preliminary Examination: |

*a. Composition:*

| *Mostly* | *Equal* | *Mostly* |
|---|---|---|
| *Mineral* | *Amounts* | *Humus* |

*b. Friability:*

*Powdery*          *Crumbly*          *Holds Together*

(Circle the terms that most accurately describe your soil sample's composition and friability, or put a check mark on the line where it belongs between two terms.)

---

2. Mineral Content:
   Sketch of a Jar at End of Settling Period, layers labeled:

---

3. Water-Holding Capacity:

*Time When Water Drained OUT =* _____

*Time When Water Poured IN =* _____

*Drainage Time =* _____

4. pH:

# Spinoff Idea

- **Effects of Leaching and Erosion on Plant Growth**

If you're wondering what makes soils different and how that might affect growing plants, here's your chance to find out more.

Soils with plenty of humus retain water and minerals, making them available to plants, but in soil with little humus, water flows through, carrying dissolved minerals with it to the ground below the root zone. That process, deep within the soil, is called leaching, and the visible effect on the surface is erosion.

Do some reading on the subject so that you know what eroded soil looks like. Then look around the school for areas of erosion. Read more about the process and do some experiments.

You'll need containers, seeds, and soil. Regular plant pots aren't necessary; you can use cardboard milk cartons just as well. Use the half-pint size, or cut the tops off larger ones (all the cartons should be about the same size). Punch holes in the bottom for drainage, and fill some cartons with soil from the eroded site and others with good garden soil. Use only one species, one whose seeds germinate readily (such as the bean), and plant one seed in each pot. You'll need a number of plants (5 to 10) growing under the same conditions before you can draw any conclusions. Remember that not all seeds germinate, so set up a few more pots than you think you'll need.

When you collect soil, be careful not to disfigure the landscape with your digging. You can return the soil when your investigation is over, but meanwhile smooth out the site to make the digging inconspicuous.

Treat all the plants the same way as far as light and water are concerned. Continue the experiment until the plants are well into the juvenile stage when they are on their own and no longer dependent on nutrients from the seeds or cotyledons. Compare the size and rate of growth of the plants grown in the two kinds of soil.

# References

American Horticultural Society. *Illustrated Encyclopedia of Gardening: Fundamentals of Gardening.* Franklin Center, PA: Franklin Library, 1982.

Botkin, Daniel B., and Edward A. Keller. *Environmental Studies: The Earth As a Living Planet.* Columbus, OH: Charles E. Merrill Publishing Co., 1982.

Foth, Henry D. *Study of Soil Science.* Chestertown, MD: LaMotte Chemical Products Co. (date unknown).

Miller, G. Tyler, Jr. *Environmental Science: An Introduction,* 2nd ed. Belmont, CA: Wadsworth Publishing Co., 1988.

# Investigation 5
# Soil Organisms

## Teacher's Section

Good growing soil is loose, not compacted, and loamy, with sand, clay, and humus particles in approximately equal amounts. Dark and damp, it is teeming with life. Each species lives in its own horizontal microhabitat within the soil, adapted to the particular conditions there: amount of moisture, light penetration, and type and amount of organic materials. Collectively, the activities that sustain the lives of soil organisms contribute to the processes of decay and soil formation.

The principal types of soil organisms and their ecological roles are listed below. The only group excluded from the list comprises the tiniest and least common of the Apterygota (wingless insects)—the proturans, diplurans, pauropods, and symphylids—which, if found, can be identified with insect guides.

## Bacteria

Soil literally teems with these microbes, among the most important of the soil organisms, and the various modes of life employed by the group enable the bacteria to fill different ecological roles. The photosynthetic bacteria live only to the depth that light penetrates, the top few inches, whereas the chemosynthetic and heterotrophic bacteria which do not require light are present throughout the soil layers. Chemosynthetic bacteria oxidize inorganic nitrogen and sulfur compounds and use the released energy for their metabolism. Most bacteria are heterotrophs, either saprophytes (living on dead organisms) or parasites (living on living organisms). A number of them are the final degraders of organic compounds, reducing them to inorganic carbon dioxide, water, ammonium, and nitrate ions while obtaining their own energy in the process.

Most bacteria are aerobic and can live wherever there is sufficient oxygen, including at depth in well-aerated soil. Anaerobic bacteria, which obtain their energy from glycolysis, cannot survive in the presence of oxygen and thus live only where oxygen is absent. Facultive anaerobes do not require oxygen and can live in its presence or absence.

The continuation of all life on earth depends on the bacteria of the nitrogen cycle. Nitrogen fixation (conversion of atmospheric nitrogen to nitrate via several steps) is carried on by symbiotic bacteria living on the roots of plants, particularly leguminous ones (*Rhizobium*, the symbiont of legumes, is the best-known example), as well as by the free-living *Azobacter* and *Clostridium*. Nitrification (conversion of ammonium ions, via several steps, to nitrate) is conducted principally by *Nitrosomonas* and *Nitrobacter*. These are the ecologically important steps, since they are the ones that make nitrate, the only form of nitrogen plants can use, available to the plants. The other step in the cycle, denitrification (reduction of nitrogen compounds to atmospheric nitrogen), is conducted mainly by *Pseudomonas*.

Other common genera of soil bacteria are the aerobes *Bacillus* and *Streptomyces*. *Bacillus* is a common rod-shaped bacteria. The filamentous *Streptomyces*, which resembles fungi, produces metabolites (metabolic byproducts) with antibiotic properties; it is also the major organism responsible for the characteristic earthy odor of soil. (The antibiotic drug streptomycin was originally derived from *Streptomyces*.)

## Algae

All algae live in water, and soil algae live in water droplets in the soil. They belong to the blue-green (Cyanophyta) and green (Chlorophyta) algal groups, both of which are photosynthetic and hence are confined to the upper layers of the soil.

## Protozoa

These organisms also are water-living, though not light-dependent. They are microscopic heterotrophs, primary consumers that feed on bacteria.

## Slime Molds

These organisms are not true molds but have some fungal properties and some animal-like ones. They are best classified as fungus-like protists. Found on damp soil, they are saprophytes, decomposers that feed on decaying vegetation.

## Fungi

All fungi are heterotrophic organisms. Some are parasitic, others saprophytic, but all obtain their nutrients by secreting digestive enzymes into the tissues of the organisms on which they feed and then absorbing the soluble digested products. Along with bacteria, fungi are the most important organisms of decay.

Among the most important fungal types are the *Mycorrhizae*, a group of mutualistic fungi associated with trees. They envelop the roots of trees, especially conifers, beeches, and oaks. Their hyphae grow into the epithelial cells of the roots and out into the soil. Minerals and products of fungal digestion and metabolism are passed into the tree's conductive system, which benefits the entire tree. Trees with *Mycorrhizae* are healthier and have a more extensive root system than those without. (Practically, young host trees are routinely inoculated with *Mycorrhizae* before transplantation, especially if they are to be planted in areas that are not naturally favorable to the fungi.)

## Nematodes

Roundworms, a major group of successful soil inhabitants, are not very well understood. Tiny, even microscopic, and free-living, they fall into two groups: the fast-moving, microbivorous worms (feeding on microorganisms) and the sluggish, herbivorous ones. The latter have a sharp stylet at the tip of the body used to pierce roots and withdraw tissue fluids. Plants often react by forming root galls, but if the attacks are severe enough, the plants may not be able to survive; nematodes are major plant killers.

## Earthworms

These important animals are scavengers that inhabit the lower, moist layers of soil. As earthworms tunnel along, they ingest soil along with whatever is in it, digesting and absorbing microbes and debris while eliminating soil, undigested materials, and metabolites as castings. The result of earthworm activity is channeling, aeration, mixing, and fertilizing of the soil.

## Snails and Slugs

These herbivorous mollusks live on the surface of the soil. Anatomically and ecologically, the two are very similar (snails have shells and slugs don't). They are slow-moving animals that glide along on a film of slime, ripping off and ingesting fragments of leaves as they go.

## Arachnids

The distinguishing characteristics of arachnids are two body parts (cephalothorax and abdomen), four pairs of legs, and the fact that they consume only liquid food. Soil arachnids include pseudoscorpions, mites, and spiders. Pseudoscorpions resemble scorpions but are smaller and lack the stinger; they are predators of insects. Mites are tiny creatures of diverse habits; some are herbivorous, while others are plant or animal parasites. The carnivorous spiders are, of course, the best-known examples of the group. They employ a variety of tactics to catch their insect prey, including webs, traps, ambush, and chase. The spiders paralyze the insects by an injection that also contains digestive enzymes. Once digestion is complete, the spider sucks its prey's body dry.

## Sow Bugs

These little arthropods, also called pillbugs or wood lice, are primitive crustaceans, one of the few terrestrial members of that group. They live in damp soil, usually under rocks or other objects, and are scavengers, feeding on decaying vegetation. When disturbed, they curl up into inanimate-looking balls.

## Centipedes and Millipedes

These two kinds of animals that superficially look so similar belong to two different arthropod classes and have quite different modes of life. Centipedes have one pair of legs on each body segment and are fast-moving predators of insects, worms, and slugs, which they paralyze by injection. Millipedes, on the other hand, have two pairs of legs per segment but despite all those legs are sluggish, remaining quietly in fallen leaves or under rocks or other objects. They are scavengers.

## Insects

The distinguishing characteristics of insects are three body parts (head, thorax, abdomen) and three pairs of legs. Insects comprise the largest and most successful group of animals on earth and live in a wide variety of habitats where they fill a variety of ecological roles. The soil is a nursery for many kinds of insects; eggs are laid in the soil and larvae develop there. Grubs (larval beetles) are commonly found in soil. The principal insect species that inhabit the soil are springtails, termites, earwigs, ground beetles, and ants.

Springtails, the most primitive of the soil insects, are small and wingless and subsist on decaying vegetation. The name comes from their means of locomotion. The last abdominal segment terminates in an extension that is tucked under the body; when this extension is released, the animal bounces foward suddenly.

Termites are also quite primitive anatomically but do have a social organization. The colony, descendents of a single queen and king, is made up of both reproductive and nonreproductive individuals. Soil-living termites construct an extensive series of tunnels through soil or through wood, eating as they go and having the same effect on the soil that earthworms do. While they are the major organisms responsible for the decay of wood, they cannot digest it without the help of mutualistic protozoa that do the actual digesting.

Earwigs are herbivores and scavengers which chew vegetation, both living and decaying. They are recognizable by a pair of projecting forceps-like structures on the end of the abdomen.

Beetles are the largest group of animals, and the ground beetles are a major family of the order Coleoptera. These beetles are fast-moving predators of other insects.

Ants, like termites, are social insects, but they are more advanced animals. They, too, tunnel in the soil, aerating and fertilizing it as they go. As a group, they are herbivores, carnivores, and scavengers. Ants have a system of social castes (queen, workers, males) and an elaborate system of communciations that enables them to follow a trail. They also manage colonies of aphids, using aphid products as a source of food.

## Mammals

The burrowing mammals, such as moles, mice, gophers, prairie dogs, and woodchucks, aerate and fertilize the soil, though on a grander scale than do the small invertebrates. Moles, secondary consumers, have well-developed adaptations for digging and spend their lives underground eating larvae, earthworms, and other soil organisms. The other animals, herbivorous rodents, are not as well adapted for digging as moles are and tend to spend more time above ground.

# *Equipment*

Several pieces of equipment are needed to obtain soil animals. They can be purchased from biological supply companies or assembled easily with ordinary materials.

## 1. Baermann Funnel

This arrangement separates nematodes from the soil. Attach a piece of tubing to the stem of a large funnel and close it with a pinchcock. Rest the funnel in a support ring, put a fine sieve in the funnel, and place a soil sample, in a cheesecloth bag, on the sieve (see Figure 2a, page 36). Any narrow-necked, broad-shouldered plastic container (for cleaning products, for example) can be converted into a funnel by cutting it off at the desired length below the shoulder. Nylon window screening ($1/16$-inch mesh) makes a good sieve.

To use the funnel, pour warm water slowly over the soil; when it is saturated, water will drip down the stem of the funnel, washing the nematodes down with it. Allow the apparatus to remain in place in the lab for about 24 hours, then open the pinchcock and drain several millimeters into a vial or dish for study.

## 2. Berlese Funnel

This apparatus, used to separate small arthropods from soil, consists of a large funnel with a wide stem and a hardware-cloth shelf (¼-inch mesh) across the middle. The funnel rests on the top of a large jar or a support ring with a jar below it. The jar is half full of 70 percent alcohol. A lamp is positioned above the funnel. (See Figure 2b, page 36.) A funnel used to drain crankcase oil (available in hardware, automotive, and discount stores) works well.

To operate the Berlese funnel, place a soil sample on the shelf and turn the light on. The soil animals adapted to cool and dark situations will move down, away from the light, and fall into the jar below. Use a low-watt bulb (25 to 40 W), and adjust the height of the lamp so it is not too close to the soil. It will take several hours, depending on the size of the sample, for the animals to collect in the jar. Then pour them into a shallow dish to examine.

### 3. Traps

A pitfall trap is sunk in the ground and catches small animals that walk across the surface. It is the easiest way to catch fast-moving animals, such as ground beetles and centipedes, that would not be taken with soil samples and might not even be seen.

A commercial pitfall trap consists of a large container with a small collecting cup inside it, and a wide-mouthed funnel that rests on the container's rim. The container is buried up to the rim (see Figure 2c, page 36). A homemade version can be made using a disposable, plastic beverage glass with sloping sides. Cut off the bottom and you have the funnel. You can combine it with any wide-mouthed jar it will fit. You can also add a collecting bottle if desired; it isn't absolutely necessary.

Attractants (bits of fruit or meat) are often helpful in luring arthropods to the trap. Killing agents usually aren't necessary, but if escape proves a problem while the trap is being removed, you can add a small amount of detergent solution or mineral oil to the jar.

Roofing is not necessary, but it does camouflage the trap from predators and interfering humans. If used, it should be a simple thing of sticks and leaves, not a more complicated construction that might alter the microclimate beneath it and thus affect which animals enter the trap.

Traps should be placed where humans will not step on or be tripped by them.

Traps can be kept going as long as you wish, but a couple of days is sufficient. If traps are in operation for more than a few days, students need to inspect them every day or so, removing the captives for examination and resetting the traps.

A potato trap is used to catch sow bugs. Hollow out the inside of a potato with an apple corer or knife and place it in a sheltered, moist area, lightly covered with leaves. In a day or two, the trap should be full of sow bugs.

## *Procedure*

The organisms of the soil make up an interactive community. This will not be obvious to students, especially if the organisms are studied apart from one another and the supporting physical environment. Educationally, it is important for students to handle the soil, discovering its various inhabitants, and to try to figure out those inhabitants' relationships.

With that in mind, divide your class into teams of three or four students and assign a site to each team. Good sites include: open areas, ones that are permanently shady, areas of mixed sun and shade, near large trees, and close to the building. The sites should be about six feet square and distributed at random on the grounds.

On-site study is primarily a matter of collection; detailed studies are done in the lab.

### Field Investigation

Have each team stake out its territory, according to previous directions (see Investigation 3), using a six-foot piece of string for measuring. Students then examine the surface of the soil, including the plants, fungi, and animals, or evidences of them.

Team members decide where they wish to take soil samples, and they should exercise the same care of the land when they dig as previously described (see Investigation 4). The most obvious soil organisms are plant roots, which should be examined to see the way they spread out. Four soil samples are obtained: one just below the surface, the second about two inches down, the third about four inches down, and the fourth about six inches down. For consistency in the class, use jars of the same size (large baby food jars are good), affix masking tape strips, and label them (1, 2, 3, 4). After each sample is put in its jar, the trowel is wiped clean to avoid mixing samples from different layers. Provide another jar for animals and other organisms found on the surface.

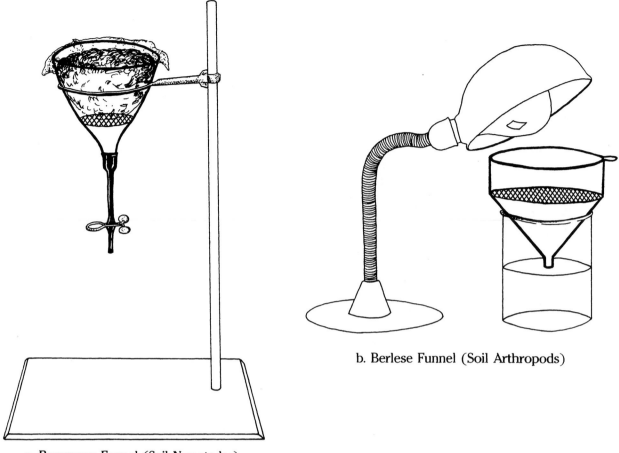

a. Baermann Funnel (Soil Nematodes)

b. Berlese Funnel (Soil Arthropods)

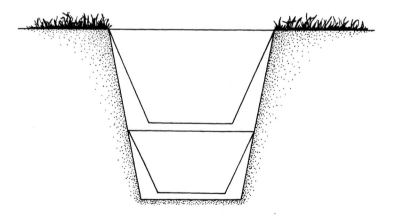

c. Pitfall Trap (Fast-Moving, Surface Arthropods)

Figure 2. Collecting Equipment for Invertebrates

After the samples have been collected, the pitfall trap is set in the hole, modifying it as necessary for a snug fit and covering it if desired.

## Lab Investigation

Students should examine the animals they find from an ecological perspective, that is, concentrating on those structures—feeding and locomotion—that determine animals' roles in the community. Magnification probably will be necessary. If some animals are too active to examine easily, especially microscopically, a short time in the refrigerator will quiet them sufficiently.

Mouthparts are specialized for biting or chewing or piercing and sucking; they determine what the animal eats, and that in turn determines its ecological role. In general, herbivores are chewers, carnivores are biters. But that is overly simplified, because some herbivores pierce plant tissues and suck fluids, while some carnivores do the same with their prey. Some animals have very specialized mouthparts (and digestive tracts) which restrict their diets; other animals have much less specialized mouthparts (and digestive tracts) and can eat a greater variety of foods.

Means of locomotion and activity levels also vary greatly. Broadly speaking, carnivores are active, fast-moving animals, whereas herbivores are less active and slower moving, and scavengers are almost sluggish. That statement applies primarily to invertebrates, not vertebrates.

### 1. LARGEST ANIMALS

Students should pick out these animals from each of the sample jars, keeping track of where they came from, examining and recording information about them. Once that has been done, the animals can be returned to the soil outside.

### 2. MICROBES

Soil microbes can be cultured on nutrient agar. Making agar plates is tedious and time-consuming, so the most efficient arrangement is to purchase ready-made plates from a supply company. They are sterile (sterilization may be difficult for you to do), and they keep for quite a long time if stored in plastic bags in the refrigerator.

Provide each team with four petri plates. A small piece of masking tape on the cover provides a place to label them (as with the sample jars). Labeling must be done without lifting the cover, and that's not easy. Students find closed containers absolutely irresistible but somehow will have to resist the urge to open them until the proper time.

When everything is organized, with jars next to their respective plates and students with freshly washed hands, inoculation can begin. The process is:

a. Open the jar and pick up a pinch of soil.
b. Lift the plate's cover with the other hand, holding it at a 45-degree angle *just* above the bottom.
c. Sprinkle the soil lightly across the surface of the agar.
d. Replace the cover quickly and tape it in place without turning the plate over.

After the plates have been inoculated, they are incubated at room temperature for several days until growth appears.

Each microbe is the beginning of a colony which is of a characteristic shape and color. When growth is well established, the shape of each colony is apparent, and the colonies are separated from one another. At that point, students should examine the plates, comparing the appearance of the different colonies with Figure 3 (page 41) and recording their findings. It is important for students to examine all the plates in the class to see the variety of colonies.

### *3. Nematodes*

The object here is for students to learn to recognize these worms and to determine their ecological roles and preferred habitat. Ideally each team has its own Baermann funnel, or several of them, but if you don't have that many, you can assign sample numbers to different teams so that all the samples are covered in the class. While this provides less material for comparisons, it may well be sufficient. If not, the extracting procedure can always be repeated.

Once students have procured the nematode-containing solution, they pour it into a shallow dish and observe the animals with a stereomicroscope. That is sufficient magnification for most of the worms, but more detail can be seen by making wet mounts.

Probably students will be able to infer the ecological role of the nematodes from their behavior: the faster-moving animals are microbivorous, the sluggish ones herbivorous.

### *4. Arthropods*

The object here, as previously, is for students to recognize these animals and to determine their roles and habitats. Arthropods, the largest and most diverse of the soil animals, are collected by various means, and each team will need to pool its collections for the study of this group. The sources of arthropods are:

a. Many of those collected on the surface of the ground.
b. Many of the large animals picked out of the soil samples.
c. Most of those extracted from the soil with the Berlese funnel.
d. Most animals captured in pitfall traps.
e. Sow bugs from potato traps.

Once students have their collections assembled, not losing track of where the animals came from in the process, they are ready to look for the devices that determine the animals' roles. Classification helps make some sense out of the collection and also gives students an idea of the enormous diversity of the group; it need only be done to the class or order or common name. References consulted for classification also provide a means of verifying student findings of feeding habits.

## *Results*

The culmination of this investigation is the construction of a food web showing the soil organisms the students discovered, their relationships, and their ecological roles. Making this web is the best way for students really to appreciate the interactions of the inhabitants in the soil ecosystem. After the web is completed, you may want to add it to your classroom wall along with the map.

## *Discussion*

You may wish to include class discussion as well. This investigation and the previous one are obviously related, and a discussion affords students the opportunity to tie the two together and help them understand the dynamic quality of the soil ecosystem. It is also a good idea to raise a few procedural questions so students can understand why they did something, rather than merely following directions. Some suggestions:

1. Why did you take all those precautions to keep the petri plates clean when you were only adding soil to them?

2. Why were different sorts of collecting and separation procedures used (Baermann and Berlese funnels, pitfall and potato traps)?
3. What role do roots play in the soil ecosystem?
4. Compare the soil organisms found at different sites and depths (that includes both similarities and differences).
5. How might the physical conditions (minerals, humus, light, air, moisture, pH) at a particular site and depth affect the different kinds of soil organisms? What are good conditions and poor ones for the different kinds of organisms?
6. If conditions, including both physical and biological ones, are favorable for a particular species, what happens? What happens if one of the conditions (your choice) becomes *slightly* unfavorable? If one or more conditions become quite unfavorable? How do these situations affect the food web?

## *Supplies for Each Team—Field*

| | |
|---|---|
| tote box | 5 baby food jars with masking tape labeling strips |
| 4 stakes, 6–8″ long, sharpened (¼″ dowel) | trowel |
| string, 6′ long | rag or paper towels |
| 6″ plastic ruler | pitfall trap |
| | hollowed-out potato |

## *Supplies for Each Team—Lab*

| | |
|---|---|
| Baermann funnel | 4 petri plates of nutrient agar with masking tape labeling strips |
| Berlese funnel | shallow dishes |
| lamp (25–40 W bulb) | |

## *Class Supplies*

| | |
|---|---|
| guides (see general reference list, pages *xiv–xvii*) | masking tape |
| bait (bits of fruit or meat) | transparent tape |
| killing agents (liquid detergent, mineral oil) | stereomicroscopes |
| | light microscopes |
| apple corer or knife | droppers and slides |

## *Spinoff Ideas*

This investigation lends itself to a number of spinoff ideas. As was true of the investigation itself, the spinoffs are centered on the organisms and their ecological relationships. Students tend to lose track of that and to see the organisms solely from the human point of view. The students may need your help to stay on course. Antibiotic activity, for example, should be seen as the action of one microbe on another, not in terms of antibiotic drugs for human diseases. The human side is considered subsequently (see Investigation 9).

Suggestions for spinoff investigations are:

1. Microbes
   a. Identification
   b. Antibiotic Activity
2. Nematodes
   a. Culture
   b. Root-Knot Formation
3. Earthworm Locomotion
4. Arthropods
   a. Feeding Behavior
   b. Daily Activity
   c. Dispersal
   d. Diversity

# *References*

Alexander, Martin. *Introduction to Soil Microbiology.* New York: John Wiley and Sons, Inc., 1977.

Behringer, Marjorie P., *Techniques and Materials in Biology.* Melbourne, FL: Krieger, 1981.

Best, Richard L. "The Pitfall Trap." *Carolina Tips,* April 1977. Carolina Biological Supply Co., Burlington, NC 27215.

Binkley, Steven W. "The Zoo Below." *Carolina Tips,* April 1984. Carolina Biological Supply Co., Burlington, NC 27215.

Hauser, Juliana T. *Techniques for Studying Bacteria and Fungi.* 1986. Carolina Biological Supply Co., Burlington, NC 27215.

Hawker, Lillian E., and Alan H. Linton, eds. "Microbiology of Soil, Air, Water," Chapter 11 in *Micro-Organisms: Function, Form, and Environment.* Baltimore: University Park Press, 1979.

Hussey, Richard S., and Ernest C. Bernard. "Soil-Inhabiting Nematodes." *American Biology Teacher,* April 1975, pp. 224–226.

Johnson, Cecil E. "The Wild World of Compost." *National Geographic Magazine,* August 1980, pp. 272–284.

Klots, Alexander B., and Elsie B. Klots. *Insects of North America.* New York: Doubleday (undated).

Kramer, David C. "Cryptozoa." *Science and Children,* March 1987, pp. 34–36.

Life Nature Libary. *The Insects.* New York: Time-Life, Inc., 1962.

Line, Les, and Lorus and Margery Milne. *Audubon Society Book of Insects.* New York: H.N. Abrams, 1983.

Milne, Lorus, and Margery Milne. *Invertebrates of North America.* New York: Doubleday (undated).

Pramer, D. *Life in the Soil.* BSCS Laboratory Block, Student Manual and Teacher Supplement. Boston: D.C. Heath and Co. (date not known).

Sherman, Irwin W., and Vilia G. Sherman. *The Invertebrates: Function and Form. A Laboratory Guide,* 2nd ed. New York: Macmillan, 1976.

Investigation 5

# Soil Organisms

## Student's Section

Now that you have determined some of the soil's important physical properties, you're ready to investigate the organisms of the soil.

Soil organisms make up a community that interacts together. The members of the community—producers, decomposers, herbivores, and carnivores—are dependent on one another for food. Your task is to find these organisms, identify them, figure out how they live, and then construct a food web to show their relationships.

Your team will be assigned to a site where you'll be collecting soil samples. Follow the same procedure you did in Investigation 4 for digging the hole, preserving soil and sod, and restoring the site after you finish.

### *Directions for Obtaining the Data—Outdoors*

#### 1. Stake-out

Stake out your site, a six-foot square. Push one of the stakes into the ground and use the string to measure where the other three stakes go.

#### 2. Observations on Site

Examine the site for living things: the plants growing there and the animals on them or on the surface of the ground. Look for evidence of unseen animals too (fur, feathers, wastes, dens). Turn over rocks, logs, or litter in your search, but do it quickly (before the animals escape), quietly, and carefully. Of course, you'll have to be on the ground yourself to do all that. Capture a few animals, put them in a jar, and examine them more closely in the lab. Record what you found on Data Sheet 1.

#### 3. Soil Collecting

Choose a spot on your site to dig for soil samples. Label four jars (1, 2, 3, 4) and collect these samples.

Jar 1—just below the surface     Jar 3—4″ below the surface
Jar 2—2″ below the surface        Jar 4—6″ below the surface

The most obvious organisms you'll see, really parts of organisms, are roots. Examine them and their extent and add your observations to Data Sheet 1. Look for fine white particles or strands in the soil; they are fungi. If you find animals while you're digging, pick them up—carefully—with your fingers or the tip of the trowel and put them in the jar with the appropriate soil sample.

After you've collected a sample, wipe the trowel clean before collecting the next. After you fill each jar, cap it and put it in a shady spot.

### 4. Traps

Set the pitfall trap in place in the hole. It should fit snugly, so you may need to enlarge the hole or pack soil around the outside of the trap. Your teacher will give you further directions about taking care of it.

Place the hollowed-out potato in a sheltered spot on your site and cover it lightly with leaves.

## *Directions for Obtaining the Data—Lab*

Now you're going to examine the soil organisms in more detail. When you examine the animals, pay particular attention to their mouthparts (biting or chewing or piercing and sucking) and try to determine what sort of food they eat. Are they herbivorous, carnivorous, or scavengers? Observe how they move and how active they are. Identify them by consulting references, verifying their eating habits at the same time.

### 1. Largest Animals

First pick out the large animals from each jar. Pour the soil into a flat dish, remove the animals, and return the sample to its jar. Mark the dish with the sample's number, identify the animals, and record your results on the proper data sheet (Data Sheet 4 for arthropods and Data Sheet 5 for other animals). Return the animals to your site when you've finished with them.

### 2. Microbes

DO NOT OPEN THE PETRI PLATES UNTIL DIRECTED TO DO SO.

Put the plates in a row with a sample jar beside each one, and label the plate covers to correspond with the jar labels. Wash your hands.

Your teacher will show you how to inoculate plates. Watch carefully, then pick up a pinch of soil in one hand, lift the cover off the plate with the other, and hold it at a 45-degree angle *just* above the bottom. Sprinkle the soil lightly over the surface of the agar and quickly replace the cover. Without turning the plate over, tape the cover in place. Repeat the procedure with the other plates. Set the plates in the designated place for several days.

Examine the plates every day and notice the microbial growth. Each microbe is the beginning of a colony, and the colonies of each species have a characteristic appearance. After a few days, when the colonies are well established, examine the plates carefully and compare them with Figure 3. Count the number of each kind of colony present and record your findings on Data Sheet 2.

### 3. Nematodes

Set up the Baermann funnel according to your teacher's directions. Once the nematodes have been separated from the soil, examine these little worms with the microscope. Observe their feeding structures and how fast they move and try to determine what they eat. Count the number of them and record your results on Data Sheet 3.

### 4. Arthropods

This is the largest and most diverse group of soil animals, so you can expect to encounter all sorts of arthropods just about everywhere. You've already collected some from the surface of the ground and as you collected soil samples. Now you'll be getting some others: with the Berlese funnel and from the traps. Combine the results of all these collections on Data Sheet 4.

Set up the Berlese funnel according to your teacher's directions. When the arthropods have been separated from the soil, pour the animals into a shallow dish to examine with the microscope as you did previously. Look for their mouthparts, try to determine what they eat, observe how they move and how active they are. Identify them by consulting references, verifying their habits at the same time.

Inspect the pitfall trap every day, removing the animals collected and resetting it. Put the animals in a jar, labeled with the date, and examine them as you did the other arthropods.

After a day or two, the potato trap should have a sizeable population of sow bugs. Count these little crustaceans and take a few back to the lab to examine as you did the other animals.

Now that the investigation is over, don't forget your site. Clean it up so it looks as if nothing happened there.

## SHAPE

Punctiform

Circular

Filamentous

Branched

Irregular

## MARGIN

Entire

Wavy

Notched

Filamentous

Branched

Irregular

## ELEVATION

Flat

Raised

Convex

Bumpy

## SURFACE

Smooth

Contoured

Figure 3. Appearance of Microbial Colonies on Agar Plates

*Biology Is Outdoors!*

# Investigation 5:
# **Soil Organisms**

| **DATA SHEET 1**—Observations on Site |
|---|
| Date: |
| Site: |
| Team Members: |
| Sketch of Site (showing where samples were dug and traps placed): |
| Plants: |
| Extent of Roots: |
| Animals Seen: |
| Evidences of Animals: |

# Investigation 5:
## Soil Organisms

| DATA SHEET 2—Microbes | | | | | | |
|---|---|---|---|---|---|---|

**Date:**

**Site:**

**Team Members:**

**Days Incubated:**

| Plate | Shape | Elevation | Surface | Margin | Color | Number |
|---|---|---|---|---|---|---|
| 1 | | | | | | |
| 2 | | | | | | |
| 3 | | | | | | |
| 4 | | | | | | |
| 5 | | | | | | |
| 6 | | | | | | |

Code:

Shape (P = punctiform, Cr = circular, Fl = filamentous, Br = branched, I = irregular)

Elevation (Ft = flat, R = raised, Cv = convex, Bp = bumpy)

Surface (S = smooth, Ct = contoured)

Margin (E = entire, Wv = wavy, N = notched, Fl = filamentous, Br = branched, I = irregular)

Color (W = white, Y = yellow, Gy = gray, Gr = green, Bl = black, Bw = brown)

Name _____   Date _____

# Investigation 5:
# **Soil Organisms**

| **DATA SHEET 3**—Nematodes | | | |
|---|---|---|---|
| Date: | | | |
| Site: | | | |
| Team Members: | | | |
| Hours Funnel Operating: | | | |

| Jar | No. Microbivorous | No. Herbivorous | Total |
|---|---|---|---|
| 1 | | | |
| 2 | | | |
| 3 | | | |
| 4 | | | |

Sketch of Nematodes:

Name _____ Date _____

# Investigation 5:
# Soil Organisms

| DATA SHEET 4—Arthropods | | | | | |
|---|---|---|---|---|---|
| Date: | | | | | |
| Site: | | | | | |
| Team Members: | | | | | |
| Hours Funnel Operating: | | | | | |
| Collection | Organism | Mouthparts | Food | Movement | Activity |
| Surface | | | | | |
| Jar 1 | | | | | |
| Jar 2 | | | | | |
| Jar 3 | | | | | |
| Jar 4 | | | | | |
| Pitfall | | | | | |
| Potato | | | | | |

Code:
   Mouthparts (B = biting, Ch = chewing, P = piercing, Sk = sucking)
   Food (C = carnivore, H = herbivore, S = scavenger)
   Movement (Cr = crawling, N = not moving, W = walking)
   Activity (+++ = very active, ++ = fairly active, + = sluggish, O = inactive)

Total Number of Arthropods by Class: _____
*Arachnida (spiders):* _____
*Crustacea (sow bugs):* _____
*Diplopoda (millipedes):* _____
*Chilopoda (centipedes):* _____
*Insecta (insects):* _____
*Total:* _____

*Biology Is Outdoors!*

Name _____ Date _____

# Investigation 5:
# Soil Organisms

| **DATA SHEET 5**—Other Organisms |
| --- |

Date:

Site:

Team Members:

| Collection | Animal | Mouthparts | Food | Movement | Activity |
| --- | --- | --- | --- | --- | --- |
| Surface | | | | | |
| Jar 1 | | | | | |
| Jar 2 | | | | | |
| Jar 3 | | | | | |
| Jar 4 | | | | | |

| Collection | Other Organisms | | | | |
| --- | --- | --- | --- | --- | --- |
| Surface | | | | | |
| Jar 1 | | | | | |
| Jar 2 | | | | | |
| Jar 3 | | | | | |
| Jar 4 | | | | | |

Code:
   Mouthparts (B = biting, Ch = chewing, P = piercing, Sk = sucking)
   Food (C = carnivore, H = herbivore, S = scavenger)
   Movement (Cr = crawling, N = not moving, W = walking)
   Activity (+++ = very active, ++ = fairly active, + = sluggish, O = inactive)

# Spinoff Ideas

## 1. Microbes

### a. IDENTIFICATION

Try to identify the microbes based on the appearance of the colonies and with the assistance of references. It may be difficult to do that because related species have similar-looking colonies, but you will be able to find the larger groups they belong to. The references will also tell you about the habits of these organisms, and from that you can figure out how they fit into the soil ecosystem.

### b. ANTIBODY ACTIVITY

Some bacteria and fungi produce metabolic byproducts (called metabolites) that diffuse out of their cells and are toxic to other microbes. You'll recognize the reaction by a ring of clear agar surrounding a colony. That is the area of highest concentration of the metabolite—so high that it inhibits the growth of other colonies. You may need to allow the petri plates to continue longer than the main investigation to see that.

Devise some experiments to study this in more detail, such as: the nature of the metabolite, the producing organism, the organisms affected by it, and the ecological role of antibiotic activity in the soil ecosystem. Focus your attention on the ecological, not the medical, effects of antibiotic substances.

## 2. Nematodes

### a. CULTURING

Place a slice of raw potato on the nutrient agar in a petri dish and add 1 ml of nematode-containing solution from the Baermann funnel. In a few days, eggs should be present and the colony growing. Observe development and feeding behavior.

### b. ROOT-KNOT FORMATION

Root knots, or galls, are caused by the reaction of plants to attack by nematodes. While the root knots are common in gardens across the country, procuring them may be difficult. Instead, obtain the name of a nematologist in the plant pathology department of your state or provincial university and write to him or her requesting a sample of root-knot infested soil for an investigation. A couple of pints is sufficient.

Sterilize garden soil by heating it in an oven at 80 to 100 degrees C (150 to 200 degrees F) for about half an hour. Punch a few holes in the bottom of several styrofoam cups or half-pint milk cartons, fill the containers with the soil, and plant tomato seeds of a variety susceptible to nematode infection (Beefeater is one). You will need six plants: three controls and three to be infected; plant a few extras. When the plants are about six inches tall, you will need to transplant them to larger containers (larger milk cartons or plant pots) for the experiment. Select six plants that look about the same for this.

First divide the nematode-infested soil among three pots, then transplant a tomato plant into each one and add sterile soil, if need be, to fill the pots. Transplant the controls to pots of sterile soil.

Examine the roots of one infected and one control plant one, three, and six weeks later. You will be able to see the galls, or enlarged areas, develop and to watch the effects on the plant, but don't be afraid of getting infected yourself. These worms only infect roots.

In order to see the worms inside the roots, the roots must be specially prepared and stained in acid-fuchsin-lactophenol solution (see references below). The chemicals used are both expensive and hazardous, so they probably are not available at your school. However, if you wish to follow this up and there is a college nearby, contact the biology department for help.

### 3. Earthworm Locomotion

Earthworms are not the easiest animals to observe in their own habitat, since they live in the lower layers of soil. However, they can be studied in our environment as long as you remember that they must be kept moist. That's moist—not soaked.

Dig up a few earthworms and watch how they move on different kinds of surfaces. Try pulling one out of its burrow. Examine the setae. From what you've seen, try to figure out the biomechanics of burrowing through the soil.

### 4. Arthropods

You could do a number of interesting investigations with these animals. For example:

#### a. FEEDING BEHAVIOR

Pick your arthropod—spiders, sow bugs, centipedes, millipedes, or insects—and watch how it eats. Notice the way the antennae are used and how food is grasped, held, and consumed, as well as any confrontations between animals over food.

#### b. DAILY ACTIVITY

Set up several pitfall traps. Collect the arthropods in the early morning and again in the early evening for a period of several days. Count the number of individuals of each species you found at each time of day. That will tell you which ones are diurnal (active during the day) and which nocturnal (active at night). Putting that information together with observations made about their feeding, locomotion, and activity (main investigation) tells you quite a bit about how the animals live.

#### c. DISPERSAL

If your area is particularly well endowed with ground beetles, you can see how they spread out in an area. Set up a large, square grid with about five traps on a side, and 10 meters (or yards) apart. This involves a lot of space and traps. Plan carefully. Plot the grid on paper and give each trap a number.

Capture 50 to 100 beetles and mark each one with a tiny dab of nail polish. Cover all the traps and then release the beetles in the middle of the grid. Uncover the traps 12 to 24 hours later and collect the animals 24 hours after that. Count the number of marked beetles in each trap, record the information by trap number, and release the beetles.

On your plot, show the number of marked beetles captured in each trap in the 24-hour period. Account for missing beetles. Study the plot to see the dispersal pattern of these animals. You can also calculate the dispersal rate, or the average distance traveled during the time period.

### d. Diversity

Set up several traps in two different habitats. After 24 hours, or longer if you prefer, pool all the captives from one habitat together and count and classify them. Do the same for the other habitat. Be sure you keep the populations from the two habitats separate. Then calculate the index of diversity.

Ecologists use the index of diversity to make comparisons between the numbers of individuals of different species found in different locations. Its formula is:

$$D = \frac{N(N-1)}{\Sigma\, n(n-1)}$$

$D$ = diversity index
$N$ = total number of individuals collected (all species)
$n$ = number of individuals of a single species
$\Sigma$ = the sum of the values for each species

You can see the calculation most easily by an example.

| Species | Habitat 1 | Habitat 2 |
|---------|-----------|-----------|
| A | 22 | 2 |
| B | 3 | 18 |
| C | 0 | 27 |
| D | 1 | 29 |
| E | 11 | 13 |
| Total | 37 | 89 |

Habitat 1:

$$D = \frac{37\,(36)}{\Sigma\, 22\,(21) + 3\,(2) + 0 + 0 + 11\,(10)}$$

$$= \frac{1332}{578}$$

$$= 2.30$$

Habitat 2:

$$D = \frac{89\,(88)}{\Sigma\, 2\,(1) + 18\,(17) + 27\,(26) + 29\,(28) + 13\,(12)}$$

$$= \frac{7832}{1978}$$

$$= 3.96$$

It is obvious that the two habitats have different numbers of inhabitants, but the diversity index provides a standardized way of comparing them.

# References

Alexander, Martin. *Introduction to Soil Microbiology.* New York: John Wiley and Sons, 1977.

Best, Richard L. "The Pitfall Trap." *Carolina Tips,* April 1977. Carolina Biological Supply Co., Burlington, NC 27215.

Brown, Vinson. *How to Follow the Adventures of Insects.* Boston: Little, Brown and Co., 1968.

Grier, James. *Biology of Animal Behavior,* 2nd ed. St. Louis: Mosby, 1989.

Hancock, Judith M. *Biology Lab Resource Book.* Portland, ME: J. Weston Walch, Publisher, 1985. (Earthworm behavior, pp. 170–173.)

Hussey, Richard S., and Ernest C. Bernard. "Soil-Inhabiting Nematodes." *American Biology Teacher,* April 1975, pp. 224–226.

Klots, Alexander B., and Elsie B. Klots. *Insects of North America.* New York: Doubleday (undated).

Life Nature Library. *The Insects.* New York: Time, Inc., 1962.

Sherman, Irwin W., and Vilia G. Sherman. *The Invertebrates: Form and Function. A Laboratory Guide,* 2nd ed. New York: Macmillan, 1976. (Earthworm locomotion, pp. 136–137; arthropod structures and responses, Exercise 7.)

# Investigation 6

# Opportunistic Species

## Teacher's Section

Opportunistic species are wild species that have taken advantage of human environments and become adapted to them. They associate with us, as they have for thousands of years, sharing common habitats and food, and often competing with us. Mostly, but not always, they cause us no harm, but because of their habits and competitiveness, we dislike them. We refer to them as weeds (plants) and pests (animals), words that reflect our values and our annoyance, not their biology. We tend to assume that if we find a species distasteful, then it should not exist. Biology courses should address that sort of thinking and those values about life and attempt to dispel them.

Technically, opportunistic species are commensals. Commensalism is a form of symbiosis in which one of the two interacting species benefits and the other is relatively unaffected. But interspecific relationships are impossible to define precisely. The natural world does not conform to our arbitrary definitions, and the different forms of symbiosis (commensalism, mutualism, and parasitism) grade into one another, blurring any distinctions we try to make. Sometimes that even happens within a species. The mosquito is a good example. The female is the obvious pest, but it is more accurately described as a predator or a parasite, not a commensal, while the male, feeding only on nectar, does not interact with us at all.

Opportunistic plants, or weeds, are herbaceous flowering plants. When compared with cultivated varieties, they are decidedly less attractive; they look "weedy," with spindly or spreading shape, coarse leaves, and inconspicuous flowers (dandelions being notable exceptions). They grow rapidly, are hardy even under adverse conditions, are highly adaptable, and have very high rates of reproduction, both sexual and asexual. These properties make them, in biological terms, enormously successful. They are also particularly well suited for their ecological role as the beginners of succession. They are the first plants to colonize barren ground, which they do from seeds or from runners spread out from already established plants. For them, open ground is a favorable habitat and, as every gardener and farmer knows, they appear wherever land is being cultivated. Many of the weeds are perennials with extensive root systems and simply amazing regenerative capacities (witness our futile efforts to dig up dandelions or crabgrass). In the winter, the shoots die back to the crowns and those lower portions of the plants survive for years.

Opportunistic animals often are the type of "wildlife" city people are most familiar with. These animals, found in the insect, bird, and mammalian groups, live in a variety of habitats and thrive on whatever food is available. Many are omnivorous scavengers, an important ecological role. Like the plants, opportunistic animals are tough, adaptable, have high rates of reproduction, and are very successful in their ecological roles.

It is not possible, of course, to provide a comprehensive list of opportunistic species, since there are so many local varieties, but the most common ones are listed in Table 1.

| TABLE 1—Opportunistic Species | | | |
|---|---|---|---|
| **PLANTS** | | **ANIMALS** | |
| Bindweed | Mallow | Ant | Rabbit |
| Burdock | Milkweed | Cockroach | Rat |
| Buttercup | Pigweed | Cricket | Sparrow |
| Chickweed | Plantain | Fly | Squirrel |
| Chicory | Poison Ivy | Gull | Starling |
| Cocklebur | Purslane | Mouse | |
| Crabgrass | Quackgrass | Pigeon | |
| Dandelion | Queen Anne's | | |
| Dock | Lace | | |
| Foxtail | Ragweed | | |
| Goldenrod | Shepherd's Purse | | |
| Hopclover | Thistle | | |
| Knotweed | Yarrow | | |

## *Procedure*

It is a good idea when you discuss opportunistic species to elicit much of the information from the class. Your students will not know the term *opportunistic species,* but they may well be able to deduce its meaning, and even the least biologically oriented among them is aware of these organisms. Thus you can begin the topic with a discussion in which all students can be expected to contribute—their prejudices, at the very least. You may need to point out some of the characteristics of these species; with skillful questioning from you, the students should be able to deduce ecological roles from their own experience and your remarks. They can certainly provide examples.

The inclination with animals is to include any species that could be described as a pest, but as you are aware from the mosquito example, not all pests are opportunistic. In order to be in this category, a species must fit the ecological description given in the opening paragraph of this investigation. Whether or not a particular species conforms makes a good point for discussion. Table 1 can serve as a reference.

If your students are like most people, they won't know individual plant names but will lump a variety of species together in the collective "weed." That is acceptable at this point; they will learn to recognize and identify local weeds in this investigation.

Organize your class into groups of two or three and assign a site to each. The sites should be at random on the lawn and in planted areas; each should be about three feet square and staked out according to the directions in Investigation 3.

Essentially, students will be looking for and identifying weeds on their sites, but they should also be alert for animals or evidence of them (feathers, feces, nests, caches of seeds, chewed vegetation). One of the characteristics of opportunistic birds and mammals is their tolerance of humans, but still these creatures are cautious. In order to have some chance of seeing the animals, students should refrain from making sudden loud noises and movements.

# *Discussion*

It is easy for this investigation to degenerate into the mundane activity of identifying weeds. In order to prevent that and keep students focused on the ecological importance of opportunistic species, it is a good idea for students to discuss their findings afterward. The intention of the discussion is to help students think beyond the obvious. Some questions:

1. What is a weed?
2. What are the ecological roles of the birds and mammals?
3. What do opportunistic species (plants and animals) have in common?
4. What roles do these species play in the ecosystem?
5. How have our activities created favorable habitats for these species? How have the species responded?
6. Many weeds have very extensive root systems that are almost impossible to dig up. What advantage is that to the plant?
7. How undesirable are opportunistic species? Should we try to get rid of them? Can we succeed?

# *Supplies for Each Group*

| | |
|---|---|
| guides (see general reference list, pages *xiv–xvii*) | string, 3′ long |
| tote box | trowel |
| 4 stakes, 6–8″, sharpened (¼″ dowel) | 4–5 stakes, about 4″, sharpened (¼″ dowel) |

# *Spinoff Ideas*

Suggestions for spinoff investigations are:

1. Growth Habits of Weeds
2. Reproductive Potential of Weeds
3. Animal Behavior
   a. Insects
   b. Flocking Behavior in Birds
   c. Mammals

# *References*

Dowden, Anne Ophelia. *Wild Green Things in the City: A Book of Weeds.* New York: Thomas Y. Crowell Co., 1972.

Fall, Michael W. "Teaching Ecology in the Urban Environment." *American Biology Teacher,* December, 1969, pp. 572–574.

Garber, Steven. *The Urban Naturalist.* New York: John Wiley and Sons, 1987.

Headstrom, Richard. *Suburban Wildlife.* Englewood Cliffs, NJ: Prentice-Hall, 1984.

Hickey, Joseph J. *Guide to Bird Watching.* New York: Dover Publications, 1975.

Klots, Alexander B., and Elsie B. Klots. *Insects of North America.* New York: Doubleday (undated).

Life Nature Library. *The Insects.* New York: Time, Inc., 1962.

Line, Les, and Lorus and Margery Milne. *Audubon Society Book of Insects.* New York: H.N. Abrams, 1983.

Slavik, Bohumil. *Wildflowers: A Color Guide to Familiar Wildflowers, Ferns and Grasses.* London: Octopus Books, 1973.

Smith, Miranda, and Anna Carr. *Garden Insect, Disease and Weed Identification Guide.* Emmaus, PA: Rodale Press, 1988.

Investigation 6

# Opportunistic Species

## Student's Section

*** CAUTION: DO NOT ATTEMPT TO HANDLE BIRDS OR MAMMALS! ***

In this investigation, you will be looking for **opportunistic species.** Opportunistic species are wild species that have taken advantage of human environments and become adapted to them. You'll be looking for both plants and animals, and both will give you some problems. The plants are easy to find, and you probably recognize a weed when you see one, but you may not know its name and so will have to refer to the guide to identify it. On the other hand, you probably know the names of the animals but will have trouble finding them. Opportunistic birds and mammals are used to humans, but they still are wary and may stay out of the way, watching you from their hiding places, especially if the class is noisy. Therefore, go outdoors and do your work as quietly as possible and don't make sudden, loud noises or movements.

## *Directions for Obtaining the Data*

WORK QUIETLY—AVOID SUDDEN NOISES AND MOVEMENTS.

First stake out your site as you did in the previous investigations. Then look it over, checking for animals and evidences of their presence (feathers, wastes, nests, caches of seeds, chewed vegetation).

Examine the area systematically for weeds, i.e., make sure you cover the whole site. Identify the weed species with the guide and note any special characteristics of their appearance. If your site is on the lawn, the plants are cut when it is mowed, so they won't look like the pictures in the guide; identify them by the appearance of the leaves. Record your results on the data sheet as you work.

After you've identified the weeds, select some of them to get rid of. Try pulling them up. If that's hard or impossible to do, try digging them up. Make every effort to remove the whole plant, but if you don't succeed, put a short marker stake in the ground next to it and check it a week or so later to see if the plant has regenerated.

While you're doing all that, watch for animals. Keep glancing up to see if you can see animals on the grounds, in the trees, or in flight. Identify them as well as you can and try to infer what sort of animals might be responsible for the evidences you found. When you actually see animals, concentrate on their behavior. Notice how they watch everything that's going on, including you; also observe their feeding, defensiveness, hiding, approaches, and interactions with other animals and people. These things tell you quite a bit about the way the animals live and give you clues about their ecological roles. Record your findings on the data sheet.

*Biology Is Outdoors!*

Name _____ Date _____

# Investigation 6:
# **Opportunistic Species**

| *DATA SHEET* |
|---|

Date:

Site:

Group Members:

Weed Species:

| Name | Appearance |
|---|---|
|  |  |
|  |  |
|  |  |

Extent of Weed Roots:

Weed Regeneration:

Ecological Role of Weeds: _____

_____

_____

Animal Species—Evidences:

| Product | Animal Responsible |
|---|---|
|  |  |
|  |  |
|  |  |

Animal Species—Seen:

| Name | Behavior | Diet |
|---|---|---|
|  |  |  |
|  |  |  |
|  |  |  |

Ecological Roles of Animals: _____

_____

_____

*Biology Is Outdoors!*

# Spinoff Ideas

What is it about these species that makes them opportunistic and able to thrive in environments we have created? To find some answers, try one of these investigations.

## 1. Growth Habits of Weeds

Select one or a few common weeds to study. Identify them, look up some information about them, and examine the growing plants to determine all you can about them: the conditions they require (soil, water, light), their growth form (prostrate, erect, creeping, climbing), height, structural details (leaves, stems, flowers, roots), and adaptations that might assist their ecological role (scent, color, runners, various means of seed dispersal).

Stake out an area that contains a number of specimens of your species and remove all of them, to the best of your ability. Check every few days to see if they reappear. Try to determine whether the new plants came from seeds or from regeneration. Continue your weeding to see if you can exterminate the weed on your site.

If, after all that, you wish to stack the deck in your favor, try a weed killer. Be sure to handle it with great care, following the package directions to the letter and avoiding getting it on anything but the weeds. Wash your hands after using it. Use controlled conditions by applying it to some specimens and not to others. Mark the plants so you know which ones you treated, when, and with how much herbicide. Compare the results. Is the plant dead if the leaves wither?

Examine regeneration under more controlled conditions. Prepare a flat or some pots with moist (not wet) soil and lay some pieces of stem or root on the soil. Cover the container with plastic wrap to keep the soil moist, keep it out of the sun, and see what happens.

## 2. Reproductive Potential of Weeds

Reproductive potential is the maximum reproductive capacity of a species. For plants, it is the average number of seeds the plants produce; for animals, it is the average number of eggs that females produce. It is not, in either case, the number of offspring that could be produced. Reproductive potential measures a species' genetic capability to reproduce, and that in turn is a measure of the ecological success of a species. It is a theoretical measurement used for comparisons, not a measure of actual reproduction.

Select a single weed with ripening seeds, identify the plant, and remove and count all the seeds. The number you get is the reproductive potential of that individual plant, not the species. You won't be able to determine the reproductive potential for the species, but you can use your seed count to make some projections.

Suppose that each seed lands on a good spot, takes root, and develops into a plant, and that each of these plants produce as many seeds (and plants) as the original one. You can keep that going for as many generations as you wish, determining the number of plants produced in each generation. In addition, you can compare reproductive potentials between different species, for example, weeds and cultivated species.

## 3. Animal Behavior

Behavior is one of the most fascinating things about animals, but it is not easy to study. Wild animals are so wary that lots of time is required just to get them used to having people in their environment before any study can begin. Domestic animals, on the other hand, are easy to observe, but their behavior has been altered by selective breeding. Thus opportunistic species which are wild but also used to people make good subjects for study. Studying behavior, however, still requires plenty of time and patience as well as the ability to make objective observations. Don't get involved if you lack those qualifications.

The following suggestions are intentionally general so you can adapt them as you wish. First select a species common to your area, then read about the behavior you're interested in. After that, you're ready to settle down and watch your animals in action.

*** CAUTION: DO NOT ATTEMPT TO HANDLE BIRDS OR MAMMALS! ***

### a. INSECTS

The opportunistic insects are quite different from one another and so lead quite different lives. Possible investigations include:

- Ants: scouting, trail laying and following, collecting food, attack and defense. Life within the colony can't be studied in natural situations but is possible with an observation cage. These cages, called "ant farms," are found in many schools, but if one isn't available, you can make one with glass or plexiglas panels attached to a bottom board. The cage should be narrow so you can see what's going on, low enough to maintain balance, and escape-proof. Fill it with soil and collect some ants for it.

- Crickets: territory, defense, male rivalry and aggression.

- Cockroaches: feeding, movements.

- Flies: feeding.

  See "References" to Hancock books for insect behavior experiments.

### b. FLOCKING BEHAVIOR IN BIRDS

A flock is a group of birds that live together in a defined territory where the feeding and nesting sites they require are available. Many flocks, especially of sea birds, are huge, numbering a million or more, while those of land birds are typically much smaller. Opportunistic birds live in flocks and provide some examples of interesting social behavior.

- Territory: advertising the territory, warnings, defense, mobbing intruders.
- Feeding: location sources, protecting the supply, communication, eating, sharing, interactions with people.
- Communications: sounds, postures, and actions used for advertisement, warning, defense, sharing information (food and danger), bickering, courtship, pair bonding.
- Courtship: courting behavior, male aggression, choosing mates, pair bonding.
- Parental Care: nest building, incubation, feeding.

### c. MAMMALS

Rabbits and squirrels are common in rural areas and parks, whereas mice and rats, even more opportunistic, are found wherever people are.

- Rabbits: feeding, alerting, defense.
- Squirrels: feeding, alerting, warnings, defense, interactions with people.
- Mice and Rats: feeding, movements. If you know of places they inhabit, you can track them and learn something about their behavior. Put some bait (bird seeds) in an area they frequent and sprinkle powdered plaster around. They walk through the plaster leaving white footprints and tailprints as they go. Examine the tracks, looking for patterns and stride length and frequency. These tell you the size of the animal and how fast it was traveling (a stride is the distance between two consecutive footfalls; frequency is the number of strides per minute). From that information, you can infer whether the animal felt safe and was moving slowly, or sensed danger and was scurrying along.

# References

Brown, Vinson. *How to Follow the Adventures of Insects.* Boston: Little, Brown and Co., 1968.

Dethier, Vincent G. *To Know A Fly.* San Francisco: Holden Day, 1962.

Fall, Michael W. "Teaching Ecology in the Urban Environment." *American Biology Teacher,* December 1969, pp. 572–574.

Grier, James. *Biology of Animal Behavior,* 2nd ed. St. Louis: Mosby, 1989.

Hancock, Judith M. *Biology Lab Resource Book.* Portland, ME: J. Weston Walch, Publisher, 1985. (Fly and cricket behavior, pp. 180–182).

Hancock, Judith M. *Project Starters for Biology Classes.* Portland, ME: J. Weston Walch, Publisher, 1988. (Ant communication, pp. 112–113; capture-recapture procedure, pp. 29–30, 49–50, 52–53; bird adaptations, pp. 118–119.)

Hickey, Joseph J. *Guide to Bird Watching.* New York: Dover Publications, 1975.

## References *(continued)*

Heintzelman, Donald S. *The Birdwatcher's Activity Book.* Harrisburg, PA: Stackpole Books, 1983.

Klots, Alexander B., and Elsie B. Klots. *Insects of North America.* New York: Doubleday (undated).

Life Nature Library. *The Insects.* New York: Time, Inc., 1962.

McFarland, David, ed. *The Oxford Companion to Animal Behavior.* New York: Oxford University Press, 1981.

Sparks, John. *Bird Behavior.* New York: Bantam Books, 1970.

Stokes, Donald W. *A Guide to Bird Behavior.* Boston: Little, Brown and Co., Vol. 1, 1979; Vol. II (with Lillian Q. Stokes), 1983; Vol. III (with Lillian Q. Stokes), 1989.

Stokes, Donald W. *A Guide to Observing Insect Lives.* Boston: Little, Brown and Co., 1983.

Stokes, Donald W., and Lillian Q. Stokes. *A Guide to Animal Tracking and Behavior.* Boston: Little, Brown and Co., 1986.

Thompson, Dale E. "The Common House Cricket." *Carolina Tips,* December 1977. Carolina Biological Supply Co., Burlington, NC 27215.

Investigation 7

# Microenvironments

## Teacher's Section

One of the delights of the natural world is discovering life in improbable places. So often we think in terms of broad sweeps of forest or field, desert or ocean, forgetting that life also exists in inconspicuous nooks and crannies, wherever it can get a toehold. There it clings tenaciously, and admirably. Truly, you can't keep the biology down!

This investigation focuses on three microenvironments found on all school grounds: those in pavement cracks, in puddles, and on shrubs.

Microenvironments, small areas with very specific physical conditions, are often transitory ones supporting life for a brief time (puddles) or to some limited size (cracks), though some are permanent (shrubs); each area supports its own miniature food chain. Generally, feeding relationships are very complex, forming an interactive arrangement best described as a food *web,* but microenvironments, lacking the diversity found in larger areas and having only a few species at each trophic level, really are food *chains.* Students can more easily see, and so understand, relationships in simpler contexts, making microenvironments an ideal subject for studying feeding relationships. Microenvironments are seldom studied, however.

The food chains of microenvironments do not necessarily contain all the trophic levels. They do, however, contain the fundamental ones found in every food web or chain: the producers (green plants) and decomposers (bacteria and fungi). Consumers are the variable group. In terrestrial microenvironments, herbaceous insects are the major primary consumers, and the top carnivores usually are spiders (secondary consumers). In puddles, algae are the principal producers, with microbes and herbaceous insect larvae the primary consumers; the chain usually ends at that point.

## *Procedure*

This investigation lends itself to variable approaches depending on the kinds of microenvironments at your school and the amount of time you wish to devote to the subject. You might decide to have all your students do all the investigations, or to choose one or two for all of them to do, or to assign different teams to different microenvironments. You will need to locate appropriate sites; size of student groups depends on the particular environment.

### 1. Life in Cracks

This microenvironment is very common, though usually overlooked. Cracks include those in pavement, in masonry walls, where the building and pavement meet, even on window sills. Wherever wind-blown particles of soil and dust accumulate and there is enough soil for seeds or spores to take root, they do, and a microenvironment is suddenly born. The plants are of variable species, but most are small or young. Moss and fungi are found in damp conditions, while flowering plants, weeds, grasses, and tree seedlings live in drier conditions; plants, then, are good indicators of soil moisture. The plants usually have herbaceous insects on them and an occasional spider.

This is a one-time investigation. Locate as many crack environments as possible and assign a pair of students to each one.

Neither a casual glance nor an erect stance will reveal this community; it is necessary to get down on its level to see what's going on. Students should just quietly observe for a few minutes, discovering the life of the crack without interfering with it. That may be sufficient to reveal all the life, but if not, students can handle the plants to look for animals. The organisms should be identified and a food chain diagram made to show their relationships.

Physical measurements are somewhat disruptive, so they are made after the biological observations. Soil depth and moisture are the critical factors for life, and they also reveal much about the fragility of this microenvironment. Students measure dimensions of their cracks, including the depth. Depth is determined by inserting a plastic ruler in the crack until it touches bottom and reading the depth directly (slice the soil first with a knife if it is too compact). Moisture is determined mainly by plant type (as noted above) and by touch. Soil that appears dry on the surface should be scraped away so the subsoil can be felt. Soil color, composition (relative amounts of minerals and humus), mineral type (relative amounts of large sand grains and fine clay particles), and pH are determined as they were in Investigation 4. Students should also decide how permanent they think these microenvironments are.

## 2. Puddle Life

Puddles are very transitory microenvironments, formed when rainwater accumulates in low spots and gone when the water evaporates or seeps into the ground. Puddle life, adapted to these circumstances with dormant stages (spores or cysts) or the ability to suspend development (larval diapause in invertebrates), is able to survive dry periods and to explode into action as soon as water becomes available. The larval stage of puddle insects is brief, often completed during the puddle's existence, with many adults having taken to wing; the rest die when the puddle dries up.

Two types of puddles form, on pavement and on soil, and their inhabitants are somewhat different as well. Life in pavement puddles, lacking contact with the soil, develops from the airborne spores of microbes and eggs and seeds of higher organisms. Mud puddles, on the other hand, have direct contact with the soil and therefore with all its nutrients and organisms. As you would expect, there is more life, and more variety, in mud puddles than in pavement puddles.

This is a good project to do immediately after a rain. Divide your class into teams of three or four members and assign each a puddle. Use both types. Students should visit their puddles as soon after the rain stops as possible and daily, or even twice daily, thereafter (depending on the climate) until the puddle dries up.

The following are determined at each visit:

### ● SHAPE

Provide each team with a ball of fairly stiff string. Students use it to outline the puddle, then measure the string's length to determine the puddle's perimeter.

### ● DEPTH

Depth is measured in the center of the puddle with a plastic ruler.

### ● TEMPERATURE

Puddle temperature is obviously related to air temperature, so students need to determine both. Water temperature is measured in the center of the puddle, air temperature nearby. Inexpensive weather thermometers are perfectly adequate instruments.

### ● PUDDLE LIFE

A meat baster is the easiest way to obtain water samples. After transferring the sample to a jar, students should determine water clarity by holding the jar up to the light and also by comparing different samples. While purely subjective, this does provide an idea of the amount of material (organic and inorganic) in suspension, and that in turn reflects the amount of life present; the cloudier the water, the more life it sustains. Animal activity can often be seen as well, but real study involves microscopy. Students should first look at their samples with the stereomicroscope to see the larger organisms and then make wet mounts to see the microbes. Puddle organisms can be identified in broad categories (bacteria, algae, fungi, protozoa, rotifers, worms, insect larvae).

## 3. Consumers on a Shrub

This one-time investigation focuses on a single plant as a microenvironment. A shrub is a large enough organism so that it creates a microenvironment in which its surfaces are more exposed to sun and wind and rain, and its interior is more protected. It is home to the insects that feed on it and to the predators, primarily spiders, that feed on the insects. Avian predators also visit the shrub but usually are not inhabitants.

A pair of students per shrub is sufficient. They should sit close to their shrub, watching quietly for several minutes, then examine it more closely, looking for insects and spiders on the stems and leaves. Bugs, beetles, and leafhoppers are the most common insects, and they don't move too far, particularly if they aren't disturbed. Flying insects can be trapped by bringing a wide-mouthed jar (peanut butter type) down over them. If it is held vertically with the open end down, they are apt to remain in it for awhile—probably long enough to be identified, certainly long enough to put the lid on. After identification, they should be released.

Spiders are the predators students are most apt to see (birds will surely avoid all the activity). The predatory tactics of spiders are fascinating and give students a good picture of the realities of life.

## *Discussion*

Discussing their results is a good way for students to make comparisons, but not all these investigations provide good discussion topics.

Cracks, wherever they are, have similar conditions, and the differences in life are primarily due to the amount of moisture present. Shrub consumers also show little variation from shrub to shrub. Neither topic lends itself well to discussion.

Puddles, on the other hand, introduce the aspect of variation, which provides the basis for comparison and hence for discussion. It is a good idea to have all the teams present their results first, grouping them by locale (pavement and mud) and then make comparisons. Some discussion questions:

1. Which type of puddle dries up more quickly? Why?
2. Which one has the clearer water? Why?
3. Make lists of the different kinds of inhabitants found in the two types of puddles. Put them on the board so they can be compared. Which one has the greater variety of life? What factors of the physical environment account for that? What happens to the living things when the puddle dries up?

## Supplies for Each Pair or Team

### 1. Life in Cracks

guides (see general reference list, pages *xiv–xvii*)
tote box
6″ plastic ruler
12″ plastic ruler

knife (may be necessary)
hand lens (useful, not necessary)
soil pH test kit
mayonnaise jar, water, detergent, teaspoon

### 2. Puddle Life

guides (see general reference list, pages *xiv–xvii*)
tote box
ball of stiff string
marking pen
6″ plastic ruler

12″ plastic ruler
weather thermometer
meat baster
baby food jar
microscopes (stereo and light)
dish, slides, dropper

### 3. Consumers on a Shrub

guides (see general reference list, pages *xiv–xvii*)

peanut butter jar

## Spinoff Ideas

Suggestions for spinoff investigations are:

1. Puddle Continuation
2. Snowmelt Puddles
3. Spider Behavior
    a. Predatory Tactics
    b. Courtship Tactics, or, How to Mate Without Getting Eaten!

## References

Baumann, Richard W. "Water Insects and Their Relatives." *American Biology Teacher*, May 1977, pp. 295–298.

Foth, Henry D. *Fundamentals of Soil Science*, 7th ed. New York: John Wiley and Sons, 1984.

# *References (continued)*

Foth, Henry D. *Study of Soil Science.* Chesterton, MD: LaMotte Chemical Products Co. (date unknown).

Klots, Alexander B., and Elsie B. Klots. *Insects of North America.* New York: Doubleday (undated).

Krall, Florence. "Mudhole Ecology." *American Biology Teacher,* September 1970, pp. 351–353.

Kramer, David C. "Cryptozoa." *Science and Children,* March 1987, pp. 34–36.

Life Nature Library. *The Insects.* New York: Time, Inc., 1962.

Line, Les, and Lorus and Margery Milne. *Audubon Society Book of Insects.* New York: H.N. Abrams, 1983.

Slavik, Bohumil. *Wildflowers: A Color Guide to Familiar Wildflowers, Ferns and Grasses.* London: Octopus Books, 1973.

Rushforth, Samuel R. "The Study of Algae." *American Biology Teacher,* May 1977, pp. 316–320.

Smith, Miranda, and Anna Carr. *Garden Insect, Disease and Weed Identification Guide.* Emmaus, PA: Rodale Press, 1988.

Investigation 7

# Microenvironments

## Student's Section

Microenvironments are small areas, separated from larger surrounding ones by distinct physical conditions that determine which kinds of organisms can live there. The biological communities are generally simple, with only a few species at each level of the food chain and often only some of the consumer levels present.

Your task is to study one or more of these microenvironments and to discover their biological communities.

## *Directions for Obtaining the Data*

### 1. Life in Cracks

With your partner, observe the life of the crack for a few minutes without interfering with it. You may be able to see all the life that way, but if not, handle the plants—carefully—while you search for animals. Identify the organisms you find and make a diagram showing their food chain relationships. Record this information on the data sheet.

Now you are ready to determine some physical properties. The most important ones are soil depth and moisture; they limit the kinds of organisms that can live in this environment. First make a sketch of the crack's shape on the data sheet, then measure its dimensions and add that information to the sketch. Determine soil depth by pushing a plastic ruler into the soil until it touches bottom and reading the depth from it. If the soil is too compact to do that, make a clean cut through it with a knife and then insert the ruler. Don't dig up the soil.

Soil moisture is determined partly by touch, but primarily by the type of plants that grow there. Moss and fungi live where the soil is damp, and that soil will feel damp to the touch. Flowering plants, such as weeds, grasses, and tree seedlings, require drier conditions. If you find them and the soil surface appears dry, scrape it away in one place so you can feel how damp the subsoil is. Replace the soil when you finish.

After that, you need to determine soil properties as you did in Investigation 4. These include soil color, composition (relative amounts of minerals and humus), mineral type (relative amounts of large sand grains and fine clay particles), and pH. Also decide how permanent you think this community is. Record all the information on the data sheet, marking the position along the line that you think best describes the physical conditions of this microenvironment.

## 2. Puddle Life

This investigation lasts for several days, beginning right after a rain and continuing until the puddles dry up. Your team will be assigned to a puddle. Visit it every day, or more often if it is drying up quickly. Determine these things on each visit and record your findings on the data sheet:

### ● *SHAPE*

Sketch the shape of the puddle on the data sheet. Uncoil the string and lay it down smoothly along the edge of the puddle. Put a mark on the string when you've gone all the way around, then pick up the string and measure its length. That is the perimeter of the puddle.

### ● *DEPTH*

Measure the depth in the center of the puddle with the plastic ruler.

### ● *TEMPERATURE*

Puddle temperature is related to air temperature, so you'll need to measure them both. Measure the water temperature in the center of the puddle and the air temperature nearby.

### ● *PUDDLE LIFE*

Use the baster to collect a water sample from the center of the puddle. Avoid stirring up the water in the process. Put the sample in the jar and determine its clarity by holding it up to the light and by comparing different samples. Look for wiggling animals at the same time.

Take your sample back to the lab to study. Pour some of the water into a shallow dish and examine it with the stereomicroscope to see the larger organisms. Identify them. Make wet mounts so you can see the microbes and identify them as well. When you finish, clean all your equipment so it is ready for the next collection.

The puddle itself changes from day to day and so do its inhabitants; you'll notice that every time you visit it.

## 3. Consumers on a Shrub

A plant the size of a shrub is large enough to create a microenvironment. Not only is it the community's producer, but also it affects the microclimate. Its surfaces

are more exposed to sun and wind and rain, while its interior is more protected. Usually there is a temperature difference between the two as well.

Sit down beside your shrub and observe it quietly for a few minutes, looking for insects, the primary consumers. Then examine the shrub more closely, looking at the stems and on and under the leaves. Do this carefully. Don't be a blizzard of activity ruffling leaves, or you'll lose what you're looking for. Watch the insects feeding and look for spiders, which prey on the insects.

Use the peanut butter jar to trap flying insects. Bring it down over an insect, hold it vertically with the open end down, and the insect is apt to remain inside at least until you can put the lid on. If you include a twig or leaf, the insect will have something to perch on, and nibble, and that makes it quieter and easier to examine. Identify the insect and release it. Determine which consumers are primary (herbivores) and which are secondary (carnivores). Record your findings on the data sheet.

Name _____ Date _____

# Investigation 7:
# **Microenvironments**

| **DATA SHEET**—Life in Cracks |
|---|
| Date: |
| Site: |
| Partner: |
| Sketch of Crack Including Its Measurements: |
| Depth: |
| Soil Color:　　　　　Light　　　　Medium　　　　Dark |
| Composition:　　　Mostly Mineral　　　Equal Amounts　　　Mostly Humus |
| Mineral Type:　　　Sand　　　　Clay |
| Amount of Moisture:　　　Wet　　　Dry |
| pH: |
| Inhabitants: |
| Community Food Chain: |
| Permanence of Microenvironment:　　　Temporary　　　Permanent |
| (On the above tests, circle the terms that most accurately describe conditions, or put a check mark on the line where it belongs between two terms.) |

Name _____ Date _____

# Investigation 7:
# **Microenvironments**

| |
|---|
| ***DATA SHEET*—Puddle Life** |
| Site: |
| Team Members: |
| Day 1—Date: |
| *Sketch:* |

| Perimeter: | Depth: |
|---|---|
| Air Temperature: | Water Temperature: |
| Clarity:  Clear  Light | Cloudy  Muddy |
| Inhabitants: | |

| |
|---|
| Day 2—Date: |
| *Sketch:* |

| Perimeter: | Depth: |
|---|---|
| Air Temperature: | Water Temperature: |
| Clarity:  Clear  Light | Cloudy  Muddy |
| Inhabitants: | |

*(continued)*

   *Biology Is Outdoors!*

## Investigation 7:
# Microenvironments

| DATA SHEET—Puddle Life *(continued)* |
|---|

**Day 3—Date:** _____

*Sketch:*

| Perimeter: | Depth: |
|---|---|
| Air Temperature: | Water Temperature: |
| Clarity:     Clear          Light | Cloudy          Muddy |
| Inhabitants: | |

**Day 4—Date:** _____

*Sketch:*

| Perimeter: | Depth: |
|---|---|
| Air Temperature: | Water Temperature: |
| Clarity:     Clear          Light | Cloudy          Muddy |
| Inhabitants: | |

Continue on another sheet if necessary.
(For "Clarity," circle the term that most accurately describes the water, or put a check mark on the line where it belongs between two terms.)

Name _____ Date _____

# Investigation 7:
# **Microenvironments**

| **DATA SHEET**—Consumers on a Shrub |
| --- |
| Date: |
| Site: |
| Partner: |
| Species of Shrub: |
| Primary Consumers: |
| Secondary Consumers: |
| Temperature at Outside Surface of Shrub: |
| Temperature at Center of Shrub: |

# Spinoff Ideas

## 1. Puddle Continuation

Instead of discarding the puddle sample after you've examined it, keep it going. That amounts to taking the puddle indoors, so keep it as much like the outside one as possible; just don't let it dry up.

Your indoor puddle should be in a wide-mouthed jar to provide enough surface area for gas exchange. Include bottom mud if you wish to. You can cover the jar with paper towel or cheesecloth (secured with a rubber band) to retard evaporation and keep dust out, or you can leave it open. However, mark the original water level with a glass marking pencil and keep the water at that level by adding more as needed. Collect rainwater or use pond or stream water, or even a little distilled water, but don't use chlorinated tap water or water from a polluted source. Keep the jar in a place where it will receive light but not be in direct sunlight.

You can keep this puddle going indefinitely. Check the inhabitants regularly and you'll be able to see completed life cycles and a succession of different kinds of organisms. You can determine food chain relationships and see how they change as the inhabitants do.

## 2. Snowmelt Puddles

Try to determine when life first appears in the spring and how much cold these organisms can tolerate. Keep track of puddle temperature from the time the snow turns to slush and meltwater accumulates and examine water samples daily, looking for the first appearance of life. Relate the different types of organisms with temperature, and you'll find a temperature-related succession.

## 3. Spider Behavior

Many people do not see spiders as interesting creatures filling an important ecological role, but as something horribly repulsive. Spiders' reputations are much worse than they should be. In fact, spiders are indifferent to people and are reluctant to bite; very few species are dangerous to us. However, since you're only watching behavior, there's hardly any chance of getting bitten.

Spider behavior is largely instinctive (almost no learning) and occurs in a definite, stereotyped way. Motionlessly, a spider lies in wait, gently swaying in its web with passing breezes, but once an insect touches a strand, the spider becomes alert, ready to spring across the web and make its kill.

Read about spiders before you begin to study them so you know something about the different species, web styles, habits, and habitats.

### a. PREDATORY TACTICS

Go in search of webs. When you find some likely-looking ones, notice the design, where and how the spider rests, its species, and its particular predatory tactics. If you have trouble seeing a web, spray it lightly and gently with water from a spray bottle. The droplets cling to the strands like dew and make the web more visible.

You might want to continue this by investigating such things as:

- Different Kinds of Insects. Are some kinds of insects caught more than others? Do spiders have preferences? Is escape possible? What about large prey?

- Can Spiders Be Fooled? **Gently** tweak a strand of the web with a feather or leaf or brush. How does the spider respond?

- Observing Web Building. Webs are damaged during a day's hunting, and each night the spider destroys the old and builds a new one, using only the sense of touch. Quite an engineering feat! Usually the new one is built in the same place as the old web, and since spiders don't object to flashlights, you can watch them work.

### *b. Courtship Tactics, or, How to Mate Without Getting Eaten!*

Generally, the male spider is much smaller than the female—rather like an insect. His problem, then, is how to approach her without having her mistake him for prey and eat him. Male tactics, which vary with the species, include:

- Waving a leg or vibrating the web in a distinct pattern from a safe distance. The response of the female to these overtures determines whether he will flee for his life or sprint across the web to her.

- Bringing her a delicacy wrapped in a packet of silk. If he is fast enough, he can mate while she opens and eats the gift. If not . . .

- Overpowering her. In those species in which males are larger, a male may be able to reach the female, tie her down with silk threads and mate with her while she is immobilized.

The life of a male spider is pretty risky, as you can see. Many are cannibalized before they get a chance to mate, and in some species, that is their fate even if they do succeed in mating. It's a tough world!

You're apt to find courting behavior by accident and may think it is predatory behavior. Check to see that the visitor at the web is also a spider, not an insect, and of the same species as the owner of the web.

## *References*

Brown, Vinson. *How to Follow the Adventures of Insects.* Boston: Little, Brown and Co., 1968.

Gray, Alice. "Eight Ways to Catch an Insect." *Science and Children,* September 1977, pp. 26–27.

Grier, James. *Biology of Animal Behavior,* 2nd ed. St. Louis: Mosby, 1989.

## References *(continued)*

Jones, Dick. *Spider: The Story of a Predator and Its Prey.* New York: Doubleday (undated).

Klots, Alexander B., and Elsie B. Klots. *Insects of North America.* New York: Doubleday (undated).

Life Nature Library. *The Insects.* New York: Time, Inc., 1962.

McFarland, David, ed. *The Oxford Companion to Animal Behavior.* New York: Oxford University Press, 1981.

Stokes, Donald W. *A Guide to Nature in Winter.* Boston: Little, Brown and Co., 1976.

Investigation 8

# Impact of the School Building on the Environment

## Teacher's Section

Any structure the size of a school building affects the surrounding environment to some extent. Some effects, such as the shadow the building casts, are obvious, but most are quite subtle. In fact, the impact that buildings have on the environment is largely an unexplored area. This investigation, a real scientific excursion into the unknown with unpredictable results, provides an unusual opportunity for students.

Data obtained in several preceding investigations form the basis for this one. The map and the following data sheets are needed for reference in this investigation:

Investigation 1—Physical Setting of the School

Investigation 3—Health of the School Grounds' Plants
   (Foundation plants, in particular)

Investigation 4—Soil Analysis
   (pH, in particular)

Investigation 5—Soil Organisms
   Microbes
   Nematodes
   Arthropods
   Other Organisms

The building's orientation (Investigation 1) is an important consideration. In the Northern Hemisphere, the prevailing winds and associated weather systems move from west to east. Thus the building's west-facing side, the windbreak side, can be expected to receive the brunt of wind and weather, though in actuality, the setting affects that markedly. Urban schools, surrounded by other buildings, are protected from the force of adverse weather, while more rural or suburban schools, located in an expanse of open fields, may be the first impediment to an advancing storm.

The surface of many buildings, both walls and roofs, is rough, a fact that has some environmental implications. As a result of weathering (the combined actions of sun, wind, precipitation, heat, freezing, and thawing), particles are continually being scraped off the surfaces. In addition, rough surfaces tend to "scrub" particulates (soot and industrial pollutants, for example) out of the atmosphere. Particles from both sources wash down the sides of the building, accumulate at the base of the walls, and leach into the soil, often in high concentrations.

The effect of these particles on the soil and hence on the life it sustains is unknown at present, according to the several soil scientists consulted. Some inferences, however, can be made based on existing knowledge. Construction materials (brick, cement, concrete, cinder block, asphalt) are made of lime, clay, sand, gravel, petroleum derivatives, and often mining slag, all substances that are affected by weathering. Lime (CaO), a major component of all such materials, provides a good example of weathering. When the surface of lime-containing products becomes wet, some of the lime reacts with the water to form hydroxide ions that become part of the washdown.

Mineral solubility depends on the pH of the soil water (Investigation 4). If enough hydroxide ions are added to the soil water from the building to raise the pH, the effect is to decrease solubility and hence availability to plants of such crucial minerals as nitrogen, phosphorus, iron, boron, copper, and zinc. Pollution particulates also contain minerals. They vary with the industries of an area, but most contain heavy metals (mercury, lead, cadmium, nickel, chromium, zinc, arsenic, and thallium), which can decrease the solubility of the biologically necessary minerals, often by forming insoluble complexes. In addition, all of these heavy metals are directly toxic to life.

Patterns of sun and shade are another way in which the building affects the surrounding area. The building's orientation on the landscape accounts for most of that, but the patterns shift with seasonal astronomical events.

The earth's axis is tilted 23.5 degrees away from the perpendicular. The axis is oriented toward the sun, or away from it, shifting that alignment during the course of a year, a fact that results in changing seasons. Summer occurs when the earth's axis is tilted toward the sun, winter when it is tilted away, and spring and fall mark the halfway points in the changing orientation of tilting. Latitude strongly affects solar illumination, and thus climate. In general, with a higher angle of incoming sunlight, solar energy is more concentrated, the climate is warmer and there are more hours of light in a day. The highest angle, 90 degrees, with the sun directly above the earth, occurs only between the tropics, and at these low latitudes there are about 12 hours of day and of night with little change throughout the year.

The other major effect of the building is that it acts as a trap for solar energy; most of the energy is absorbed by masonry, while the rest radiates back to the environment. Also, building heat passes outward, dissipated through the walls and roof, to the environment. As a result of these thermal activities, the soil adjacent to the building is warmer than its surroundings, with a temperature gradient that spreads outward from that high point.

The building also acts as a barrier. Windblown materials of all sorts crash into the walls or flutter against them and come to earth at their bases. Some are noticeable, others not, but any of them can leach chemicals into the soil and so affect the environment in ways beyond the merely aesthetic.

# *Procedure*

The first parts of this investigation are concerned with the physical influence of the building on the environment, setting the stage for the last part: its influence on life. Students may find some answers to the questions they asked—or wondered about—in connection with earlier investigations.

## 1. Evidences of Weathering

A simple demonstration is a good way to begin this investigation. Visit a construction site and pick up a few small pieces of cement. The raw material, a mixture of baked lime and clay, is combined with sand and water to make a slurry that is used to cement masonry products together. It dries to rock-hard consistency.

Put a piece of the hardened cement in a small dish and pour acid solution over it. Oxygen bubbles appear immediately. Leave the cement in the solution for several days, or even weeks, and the bubbling disappears and the solution gradually becomes discolored as sand and clay particles accumulate on the bottom of the dish. These things are indicative of chemical changes.

Any acid can be used (masons use "muriatic acid," hydrochloric acid, to remove excess cement from brickwork). You can use this, but probably the easiest to use is vinegar (5% acetic acid), which can be used directly from the bottle. It is a weak acid, with a pH (about 4) in the same range as rain in the eastern part of North America, so in addition to being a convenient agent for the demonstration, it also simulates a natural situation. (Uncontaminated rain is also acidic, pH 5.6; lower values are considered "acid rain.")

Test the pH of the acid solution before and after adding the cement, and finally after the solution has become discolored. In my preliminary tests, the pH of the vinegar (4) was unchanged by adding the cement, but the tan solution several weeks old had a pH of 6. That increase is due to the gradual accumulation of hydroxide ions as the lime became hydrated.

Students may doubt that chemical changes can occur in building materials, and this demonstration is more convincing than words. It also prepares the students to see that life could be affected as a result.

The investigation itself is a one-time activity involving the whole class.

The extent of weathering varies greatly with climate, location, and air pollutants, but all structures are affected. The amount of time the building has been exposed to the weather is also a factor, so find out from the office when the school was built.

Take your class outdoors to look for evidence, beginning at the west-facing side and progressing around to the north, east, and south sides. Refer to the orientation of the building (Investigation 1 or the map).

The evidences to look for are:

- Rough surfaces that lose particles when rubbed by hand.
- Accumulations of sand and gravel at the base of walls and under the drip line of overhanging roofs.
- Flakes of paint on the ground and flaking paint on window and door trims.
- Discharge from downspouts.
- Windblown debris stopped by the building, as evidence of wind action.

## 2. Sun/Shade Patterns

This study also depends on the building's orientation. From that information, students can track the sun's course across the sky from rising to setting. The positions of the rising and setting sun, with respect to the horizon, change with the season, reflecting the angle of incident rays meeting the tilting earth, a fact that need not concern you here.

To investigate the sun/shade patterns at your school, take your class out on a good clear day. The class can work as a whole, initially. Students should first note the time and estimate the sun's position relative to the horizon, then walk around the building, noting which sides are entirely exposed to sun, which are shaded, which have both sun and shade. The length of shadows should be measured.

As it stands, that information is meaningless. In order to cover the daylight hours more completely, divide your class into teams and assign each an hour to check sun/shade patterns. You needn't include all hours, but do try to include the critical points: early morning, noon, and late afternoon.

The data can then be plotted to show shifting sun/shade patterns. It is important that the data be comparable. The best way to insure that is to copy one of the outline drawings of the building (Investigation 1 or the map), affix it to the data sheet, and photocopy the whole thing. Students can then simply draw in the shady areas and add measurements. When the data sheets for different times of the day are put together, the moving patterns of sun and shade become apparent.

## 3. Influences on Soil and Soil Organisms

When weathering products and debris reach the ground, they accumulate; soluble components dissolve and wash downward, while insoluble ones remain on the surface. The highest concentration of these materials is in the washdown areas: at the base of the walls, under the roof's drip line, and channeled below downspouts, with a concentration gradient spreading outward from those points. Heat energy also radiates outward in a gradient from its high point, the building.

Select several sites along the sides of the building, divide your class into teams of three to four members, and assign each a site.

The gradients to be determined are soil temperature and pH. These determinations are made at regular intervals along a line perpendicular to the building. To do that, provide each team with two small stakes and a piece of string about six feet long. Students tie the ends of the string to the stakes, push one stake into the ground at the base of the building, then run the line out so it is taut and at right angles to the building, and drive in the second stake.

The following determinations are made at the base of the wall and at six-inch intervals along the test line until two consecutive readings are the same. The gradient high points mark the peak temperature and pH readings (either high or low); the levels gradually fall away from there.

### a. TEMPERATURE

Use soil thermometers. Comparable in cost to full-range lab thermometers, these instruments, obtainable from garden or biological supply companies, have protective metal shields over the stems to avoid the hazards of breakage and loss of mercury, and so they are safer and have longer lives than the usual thermometers. Soil temperature is measured at a depth of five to six inches at each position along the test line and the gradient high point determined.

### b. pH

Soil pH is measured with the soil test kits at the same depths and positions as the temperature, and the gradient high point is also determined. These results can be compared with those obtained in Investigation 4.

### c. SOIL ORGANISMS

This part of the investigation is not very different from Investigation 5. However, the intention here is to see whether temperature and pH deviations might affect soil organisms. Investigation 5 is the comparison group for this investigation.

The position of teams along the sides of the building and of testing positions along the line are such that little variation in the living community is expected. The gradient high points are of particular interest. Once a team has determined those points, members can examine the soil there looking for larger soil organisms, which can be identified and counted and compared with those found in Investigation 5. The effects are expected to be most pronounced for the smallest organisms, the microbes and nematodes. You can limit the investigation to them and have alternate teams do one or the other, following the procedures in Investigation 5. The critical thing to look for is the number of individuals of different species present, as compared with Investigation 5.

### d. Foundation Plants

Students should reexamine their data on the health of foundation plants (Investigation 3), or the plants themselves, from the perspective of this investigation. The position of these plants relative to the building and the gradient high points is the important consideration.

## Discussion

The role of a building in its environment has not been investigated by scientists, to the best of my knowledge. In the absence of knowledge, it is not possible to describe what results, if any, will be found, but it is expected that they will vary tremendously depending on location. Atmospheric pollutants from different industries, including acid rain, can accelerate weathering and/or mineral solubility. It is altogether likely for students in some locations to find pronounced effects while those in other places find no effects or only vague ones.

Students often think in terms of dramatic results and feel faintly cheated with anything less. But the natural world is not like that. It is far more subtle than dramatic and can offset or buffer—within limits—many of the hazards that befall it. Discerning subtle effects requires very knowledgeable people working with very sophisticated equipment, neither of which are found in the classroom. Still, that should present no educational barrier.

The whole point of investigation as an educational tool is to give students an opportunity to stretch their minds by trying to solve problems about the unknown. They need discussion to learn from one another: to tie their findings together, to speculate about possible causes, and to express their ideas and tolerate different ones. They need your help to realize that there are limits: to knowledge, to their present ability to acquire it, to the natural world's buffering capabilities. This investigation should provide that sort of fundamental education very well.

For the discussion, students should have available for reference the data sheets for this investigation as well as those for the earlier ones listed at the beginning of this investigation (and on the student pages). They should also be conversant with the critical terms: *weathering, leaching, gradients, pollutants.*

A good way to launch the discussion is with a couple of broad-based questions:

1. How did you expect the building to affect the environment?
2. How were your results different from your expectations?

After a few minutes of general discussion, begin to shift toward the specific. Have the class list these data on the board for reference:

1. A list of weathering evidences and products found.
2. Temperature and pH determinations, arranged by site and position, with gradient high points marked.

Provide the class with a third list, of typical air pollutants, most of which come from these six classes of substances:

- carbon oxides
- sulfur oxides
- nitrogen oxides
- volatile organic compounds (hydrocarbons and derivatives such as formaldehyde, a common component of synthetics)
- particulates (dust, soot, heavy metals, asbestos, and droplets of acid, oil, and pesticides)
- photochemical products, i.e., substances formed when sun energy causes two or more of the other components to react (ozone, acids, and peroxyacyl nitrates, or PANs)

Pollutants, by definition, are substances, natural or synthetic, that are concentrated to the point where they cause harm to living things or to inorganic materials, such as building materials. We understand rather little of the chemistry involved, or of the effects.

More specific discussion questions:

1. How does the slope of the land (Investigation 1) affect the distribution of weathering and "scrubbing" particles through the soil? What happens where there is no slope?
2. How do variable sun/shade patterns affect the soil and its inhabitants?
3. How can you account for the gradient high points? To what extent do you think the building is responsible?
4. Did the gradients seem to cause any effects on soil organisms? (Compare with Investigation 5.)
5. To what extent do the building and/or the gradients affect the health of foundation plants? Look for both positive and negative effects.
6. How much impact do you think your school has on its environment?
7. Could a building's impact be reduced? How?

## *Class Supplies*

| | |
|---|---|
| guides (see general reference list, pages *xiv–xvii*) | Baermann funnel |
| small pieces of cement | nutrient agar petri plates, with masking tape labeling strips |
| small dish | stereomicroscopes |
| vinegar (or other dilute acid) | light microscopes |
| pH test paper | Berlese funnel and light |
| map (from Investigation 1) | shallow dishes, droppers, and slides |
| carpenter's tapes, 8′ or longer | |

## Supplies for Each Team

| | |
|---|---|
| tote box | soil thermometer |
| 2 stakes, 6–8″ long, sharpened (¼″ dowel) | soil pH test kit |
| string, 6′ long | trowel |
| ruler | baby food jars with masking tape labeling strips |

## Student-Supplied

data sheets for Investigations 1, 3, 4, 5

## Spinoff Ideas

Suggestions for spinoff investigations are:

1. Seasonal Effects on Sun/Shade Patterns
2. Ecochemistry of Weathering
    a. Wood Preservatives
    b. Acid Rain
    c. Effects on Soil Organisms

## References

American Horticultural Society. *Illustrated Encyclopedia of Gardening: Fundamentals of Gardening.* Franklin Center, PA: Franklin Library, 1982.

Bienz, D.R. *The How and Why of Home Horticulture.* San Francisco: W.H. Freeman Co., 1980.

Botkin, Daniel B., and Edward A. Keller. *Environmental Studies: The Earth As a Living Planet.* Columbus, OH: Charles E. Merrill Publishing Co., 1982.

Foth, Henry D. *Fundamentals of Soil Science,* 7th ed. New York: John Wiley and Sons, 1984.

Foth, Henry D. *Study of Soil Science.* Chestertown, MD: LaMotte Chemical Products Co. (date unknown).

Hawker, Lillian E., and Alan H. Linton, eds. "Microbiology of Soil, Air, Water," Chapter 11 in *Micro-Organisms: Function, Form and Environment.* Baltimore: University Park Press, 1979.

Hussey, Richard S., and Ernest C. Bernard. "Soil-Inhabiting Nematodes." *American Biology Teacher,* April 1975, pp. 224–226.

Kramer, David C. "Cryptozoa." *Science and Children,* March 1987, pp. 34–36.

Nadel, Ira Bruce, and Cornelia Hahn Oberlander. *Trees in the City.* New York: Pergamon Press, 1977.

Pramer, D. *Life in the Soil.* BSCS Laboratory Block, Student Manual and Teacher Supplement. Boston: D.C. Heath and Co. (date unavailable).

Investigation 8

# Impact of the School Building on the Environment

## Student's Section

The subject of this investigation is one you may never have thought about before: how a building can influence its environment. If you haven't, you're not alone, because this is something almost no one thinks about, scientists included. So there isn't a lot known, and you may truly discover something new in your investigation.

You'll be using some of the data you've already collected and a number of the same procedures, along with some new ones, but now you'll be seeing how all that information fits together and trying to find some causes. In the course of this investigation, you may find answers to some of the questions you had earlier; you'll find more questions, too.

You'll need these data sheets:

Investigation 1—Physical Setting of the School

Investigation 3—Health of the School Grounds' Plants
  (Foundation plants, in particular)

Investigation 4—Soil Analysis
  (pH, in particular)

Investigation 5—Soil Organisms
  Microbes
  Nematodes
  Arthropods
  Other Organisms

# Directions for Obtaining the Data

## 1. Evidences of Weathering

The extent of weathering varies greatly with the location and the amount of shelter the building has, but all construction materials weather during the course of time. Here are some evidences of weathering to look for:

- Rough surfaces that lose particles when rubbed. Run your hand across the surface, then look for particles or cement dust on your hand.

- Accumulations of sand and gravel at the base of walls and under the drip line of overhanging roofs.

- Flakes of paint on the ground and flaking paint on window and door trims.

- Discharge from downspouts.

- Windblown debris that was stopped by the building. This is not the result of weathering, but it does indicate wind action.

Record your results on the data sheet.

## 2. Sun/Shade Patterns

Sunlight, as you know, is a biological necessity, but as you also know, it isn't always available. Here you will be investigating the building as a block to sunlight. When sunlight strikes one side of the building, it casts a shadow on the opposite side, but because of the earth's spinning and orbiting around the sun, sun and shade patterns change throughout the day, as well as by seasons.

Your task is to determine these moving sun/shade patterns over the course of a few days. First check the building's orientation on the map. Once outside, note the time and the sun's position relative to the horizon, then walk around the building to see the sun and shade patterns. Draw the shade pattern on the data sheet, measure its extent and record that along the sides of the shade pattern.

Different teams will check the patterns again at different times. When all the teams' results are put together, you'll be able to see the moving pattern of sun and shade around the building. That's only one brief time, of course; it takes a full year to get the complete picture, but you don't have to do that!

## 3. Influence on Soil and Soil Organisms

You're going to be doing tests of the physical environment to determine if gradients exist. A gradient is a gradation in the concentration of substances or energy from one place to another as a result of diffusion. The particular properties you'll be concerned with are temperature and pH.

In order to determine whether gradients are present, you make measurements at regular intervals along a test line at right angles to the wall of the building. First tie the ends of the string to the stakes. Then drive one stake into the ground at the base of the building, run the line out so it is taut and perpendicular to the building, and drive in the second stake. Do the tests at the base of the building (position 1) and at six-inch intervals along the test line until two consecutive readings are the same.

Record your results on the data sheet and determine the gradient high points, that is, the peak temperature and pH readings (either high or low) from which the other readings grade away.

### a. TEMPERATURE

Put the stem of the thermometer in the ground to a depth of five to six inches. Wait for the mercury to reach temperature (several minutes), take the reading, and move on to the next position.

### b. pH

Take the soil sample for this test at the same depths and positions as the temperature. The procedure is the same that you used in Investigation 4.

### c. SOIL ORGANISMS

This is not just the same as Investigation 5, because here you are interested in seeing whether the high points of the gradients have any effect on the populations of soil organisms. Your team will probably only be looking for certain kinds of organisms, and you'll find out about others from the results of the other teams. The directions for obtaining soil organisms are given in Investigation 5. Compare the results you get here with those in Investigation 5 to see if there are any differences.

### d. FOUNDATION PLANTS

Reexamine your data on the health of foundation plants (Investigation 3), or look at the plants again. See if there is any relationship between the health of the plants and their position in relation to the building and the gradient high points. Record your observations on the data sheet.

Name _____ Date _____

# Investigation 8:
# Impact of the School Building
# on the Environment

| **DATA SHEET**—Evidences of Weathering |
| :--- |
| Date: |
| Year School Was Built: |
| Construction Materials (from Investigation 1): |
| Evidences of Weathering: |
| Part of Building Most Affected by Weathering: |
| Overall Extent of Weathering:<br>    None        Very Little        Some        Extensive<br>(Circle the term that best describes the extent of weathering, or put a check mark on the line where it belongs between two terms.) |

Name _____ Date _____

# Investigation 8:
# Impact of the School Building
# on the Environment

| **DATA SHEET**—Sun/Shade Patterns |
|---|
| Date: |
| Time: |
| Team Members: |
| The Sun's Position Relative to the Horizon: |
| W_____E |
| Patterns of Shade, Including Their Measurements: |

# Investigation 8:
## Impact of the School Building on the Environment

| **DATA SHEET**—Influence of Soil and Soil Organisms |
| --- |

Date:

Site:

Team Members:

|  | Positions | | | | | |
| --- | --- | --- | --- | --- | --- | --- |
|  | 1 | 2 | 3 | 4 | 5 | 6 |
| Temperature: |  |  |  |  |  |  |
| pH: |  |  |  |  |  |  |

Gradient High Points:

*Temperature:* _____

Reading: _____

| Position: | Distance from Building: |
| --- | --- |

*pH:* _____

Reading: _____

| Position: | Distance from Building: |
| --- | --- |

Soil Organisms: _____

*Species:* _____

*Position on Gradient:*




*Numbers:* _____

Foundation Plants:

*Species:* _____

*Position on Gradient:*




*Health:*

# *Spinoff Ideas*

## 1. Seasonal Effects on Sun/Shade Patterns

In order to determine the sun/shade patterns at your school for all four seasons, you will need some basic information. The person to consult is a local meteorologist. You can find one in a TV station's news department. Call for an appointment, briefly explaining what you're looking for, and then you can meet to discuss it more fully. The information you need is:

- The latitude of your area.

- The compass positions of the rising and setting sun for your area for the longest and shortest days of the year.

- The length of the longest and shortest days.

- The angle of the sun's rays to earth in your area at noon on the longest and shortest days.

- The average number of sunny days for the year.

The meteorologist may give you the information directly or may suggest references where you can find it yourself.

Once you have that information, you can track the sun's position across the sky on the first day of summer and of winter and figure out its position on the first day of spring and of fall (halfway between the other two). From that and the school's orientation, you can determine sun/shade patterns at different seasons and times of the day. Which areas around the school receive the most sun in the course of the year? Which the least? Based on your study, what planting recommendations would you make?

## 2. Ecochemistry of Weathering

Several aspects of this subject are interesting to follow up. If you decide to investigate one of them, keep in mind that the focal point of this course is on life, not on inanimate materials, so you'll need to center your investigation on life (see below).

### a. WOOD PRESERVATIVES

A variety of substances are used to prevent wood from rotting, such as creosote, pressure treatment, paint, and stain. You need some information about their chemistry and what happens when they weather. That may require several sources: an encyclopedia for basic information, a chemist, a technology professor at a university (i.e., someone who knows the theoretical aspects, rather than a builder). How might weathering products, or spills, affect soil organisms?

### b. ACID RAIN

If acid rain is a problem where you live, you might want to investigate it. Consult a local meteorologist (see above) to learn the pH of normal and acid rain in your area. Then check references to learn the effects of acid rain on construction materials and living things, and experiment with soil organisms.

### c. EFFECTS ON SOIL ORGANISMS

Microbes are the best organisms to use for experiments; they are the easiest to collect and work with, and they are the most important of the soil organisms. Follow the technique given in Investigation 5 and choose one of the following procedures. Put the experimental material directly on the agar, then add the soil sample and culture the plates as you did in Investigation 5.

- Use acid or basic solutions of different and known pH. If you have not studied chemistry, you do not know how to make these solutions, so don't try to do it. Consult your teacher. Solutions in dropping bottles are the easiest to use, since you don't need very much. Test each solution with the pH test paper.

  Put a few drops of one solution on the surface of the agar, replace the cover, and gently rock the petri plate to distribute the solution. Then add the soil sample. Set up a graded series of plates using progressively stronger solutions (increased pH for bases, decreased pH for acids). Label the plates.

  After a few days, check the plates and compare the results, both within the series and with those you got in Investigation 5, and compare the species present with Figure 3. Are there any differences?

  \*\*\* CAUTION: DO NOT COMBINE SOLUTIONS! \*\*\*

- Use a few drops of creosote, or paint, or stain, or tiny slivers of pressure-treated wood, spreading the solution and treating the plates as described above.

  \*\*\* CAUTION: DO NOT COMBINE MATERIALS! \*\*\*

You might want to investigate more than one of these variables, but do only one of them at a time. Combinations of chemicals give uncertain results—you don't know which one is responsible for what, or they may cancel one another out. Large quantities might be dangerous.

# References

Alexander, Martin. *Introduction to Soil Microbiology.* New York: John Wiley and Sons, 1977.

Angyal, Jennifer. "Acid Rain: The Bitter Dilemma." *Carolina Tips,* September 1980. Carolina Biological Supply Co., Burlington, NC 27215.

Botkin, Daniel B., and Edward A. Keller. *Environmental Studies: The Earth As a Living Planet.* Columbus, OH: Charles E. Merrill Publishing Co., 1982.

Hawker, Lillian E., and Alan H. Linton, eds. "Microbiology of Soil, Air, Water," Chapter 11 in *Micro-Organisms: Functions, Form and Environment.* Baltimore: University Park Press, 1979.

## References *(continued)*

LaBastille, Ann. "Acid Rain: How Great a Menace?" *National Geographic Magazine,* November 1981, pp. 652–680.

Miller, G. Tyler. *Environmental Science: An Introduction,* 2nd ed. Belmont, CA: Wadsworth Publishing Co., 1986.

Investigation 9

# Impact of People
# on the Environment

## Teacher's Section

The impacts that people exert on the environment are many and varied, ranging from constructions of all sorts to agriculture, commerce, and recreation. All human activities have some impact on the environment.

In the case of a school, a major factor is the building itself (Investigation 8). The surrounding grounds, whether paved or landscaped, constitute an artificial environment that favors cultivated plants (Investigation 2) and opportunistic species (Investigation 6) at the expense of native species. In urban settings, most native species have long since been lost, though pockets may remain in parks, while in rural or suburban locations, native plants and animals are often found in the vicinity of the school and there, too, natural areas are frequently preserved in campus-style settings (Investigation 10).

The major environmental impacts at a school are:

## ● Noise

A great many people make a great deal of noise, particularly if most of them are teenagers! That's enough to keep the vast majority of reptiles, birds, and mammals at a distance. Generally, invertebrates and lower vertebrates do not have as acute sensory perception and therefore are less affected by human noise. Opportunistic birds and mammals, possibly not as sensitive as most, have adapted to the noise and carry on their lives normally in the presence of people.

## ● Traffic

Many feet wear the vegetation down and compact the soil. As a result, water and air do not penetrate, plants die, humus is washed away, and gradually a path, of the off-white shade of native minerals, is worn. There is no soil. Depending on the depth of compaction, soil organisms are killed, rendering the path sterile. Since high-traffic areas are barren and none too attractive, many of them are paved, further compacting the ground and often killing tree roots that pass underneath.

## ● Refuse

Wherever there are people, there is trash. Pieces of refuse often become shelters for small animals or microenvironments (Investigation 7), supporting communities of life that decompose organic materials. Most synthetics, especially plastics, resist biological decomposition and last for years, removed from the area only by cleanup. Refuse may create a microclimate under it that favors certain organisms (fungi in damp and shady spots, for example).

## ● Construction Sites

These are places of active maximum environmental disruption: noise and human activity abound; trees and bushes are cut; soil ("topsoil") is scraped up, destroying microhabitats in the process; the more sensient animals take flight, and a great many of the small ones are killed. It is a natural disaster. But there are survivors, and once the activity is over and the soil respread, they begin to reconstitute and repopulate the microhabitats.

# *Procedure*

The previous investigations all required you to provide some background information before students could begin their investigation. But this one is different. Students are aware of, or can easily deduce, the impacts that people have on the environment. Discussion is the best way to begin this topic.

From a practical standpoint, too, the subject is not as easy to investigate as previous ones, primarily due to the lack of contrasting conditions. You cannot, for example, eliminate school noise, nor can you get native species back into an environment that no longer has the habitats they require. Some suggestions for investigation are included, following the discussion questions, but you might also wish to elicit ideas from your class based on specific conditions at your school.

When you begin the discussion, it is a good idea to let it be quite general at first, gradually steering it toward the specifics of your school. For example:

1. What is meant by *impact*? Does the word have only a negative connotation, or does it have a positive one as well?
2. In general, how do people exert an impact on their environment? (List on the board.)
3. What are the particular impacts of people on a school environment? (List on the board.)
4. Give examples of environmental impacts at your school. (List on the board.)
5. Which environmental impacts would be interesting and possible to investigate? How would you go about investigating them? Devise some procedures.

These suggestions for investigations are based on the kinds of human impact discussed above.

## 1. Noise

In scientific studies, sound levels are recorded on instruments that measure them in decibels, thus providing a numerical basis for comparisons among different conditions. However, these instruments are expensive and schools usually cannot justify their cost.

A simpler, though perfectly satisfactory, arrangement is to use a tape recorder. Recordings should be made outside the building at different times during the day to provide a spectrum of typical school sounds. A contrasting recording can be made, if desired, in a natural area.

Playing the tapes in class can provide the basis for a discussion (see below) on the effects of sound on animals. Some students may wish to follow that up further; information is available on the subject.

The obvious experiment, playing the school tape in a natural area, can also be done. For it, students will need some guidance in the art of quietness. Their inclination is to play the tape at maximum volume, but subtlety is in order. A student or two (not more) should go into the area, find a comfortable spot, and wait quietly for the resumption of natural sounds. Birds are the critical indicators. They stop singing, or issue warning calls, if disturbed, so students must listen carefully to the songs, then play the tape at its lowest volume, noting any changes in the singing. If it stops altogether, then students can stop the recording and wait until the singing begins again. Should the lowest volume eliminate some songs, but not others, the volume can be raised slightly and gradually until all singing stops, then stopped or resumed to stimulate the birds' responses. This activity, which requires patience and good listening skills, rewards the student with the feeling of having communicated with other creatures.

## 2. Effect of Soil Compaction on Soil Organisms

Students choose a well-worn path and take soil samples as they did in Investigation 5. If the samples are taken at the same depths (immediately below the surface and two, four, and six inches down) and examined for the same organisms (microbes, nematodes, arthropods, and others), the results can be compared directly with those obtained in Investigation 5, which then becomes the control group.

## 3. Refuse

Good-sized pieces of refuse, particularly containers, may become shelters or microenvironments. If they act as heat traps, they create microclimates, in which the temperature and moisture within or under them is somewhat higher than the surroundings and the light less. You can determine these properties easily with a thermometer, with an inexpensive humidity meter, and by noticing shadows. If you don't have a humidity meter, you can get an approximation of the comparative humidity (between the microenvironment and its surroundings) by using drying crystals (calcium chloride) to pull moisture out of the air, or even a cold glass on which atmospheric moisture can condense.

Pieces of refuse should be examined for animals by opening or shaking them over a container and also by looking for evidence, such as chewed edges.

## 4. Construction Sites

If construction work is going on at your school, you might wish to visit the site so students can discover environmental impacts for themselves. The most striking aspect of any construction site, besides the devastation, is the expanse of exposed, barren subsoil. This gives students a chance to see what the underlying minerals look like en masse and to appreciate, better than any other way, the living and life-giving quality of soil. Often the soil is preserved, piled up in huge heaps where its dark color makes a dramatic contrast with the parched, off-white shade of the subsoil.

## *Follow-up Discussion Questions*

Some discussion questions are given here. The first four are based on the preceding suggestions for investigation and may need to be modified to suit your particular situation. The last question is very broad and should be included regardless of which investigations your class did because it gives students a chance to discover some basic truths about people and the environment: that environmental impacts are a fact of life, that we cannot have the institutions of society without some effect on the environment, and that the role of a school is to serve young people, who by their very nature are hard on the environment. Students need to learn about priorities and that trade-offs are part of human reality. Desirable as protecting the environment is, education is also important, and here, in this place, education must take precedence over environment. At the same time, the environment should be cared for to the extent possible. These are things that students should learn as they become aware of ecological realities.

1. What are the responses of wild animals to loud noises in their environment? (Tapes, personal experiences, information from reading, and inferences will add a great deal to this discussion.)

2. What are the effects of compaction on soil organisms? What physical soil properties are affected? How could compaction be reversed?

3. What are the climatic effects of refuse? How do living things take advantage of the refuse you provide?

4. What sorts of environmental impacts are created by construction work?

5. How much environmental impact do the people who use the school create? To what extent are these impacts detrimental? beneficial? Could you eliminate environmental impacts? Which ones? How?

## *Class Supplies*

| | |
|---|---|
| recorder and tapes | Berlese funnel and lamp |
| trowel | weather thermometer |
| 6″ plastic ruler | humidity meter, or drying crystals, or cold glass |
| baby food jars with masking tape labeling strips | bowl or pan |
| Baermann funnel | |

## *Student-Supplied*

data sheets from Investigations 4, 5, 6, 7

## *Spinoff Ideas*

No spinoffs are included. The suggestions for investigation offer some possibilities, and students can think of others that pertain particularly to their school. The spinoff in Investigation 10 is also applicable to this one.

## *References*

Botkin, Daniel B., and Edward A. Keller. *Environmental Studies: The Earth As a Living Planet.* Columbus, OH: Charles E. Merrill Publishing Co., 1982.

Kramer, David C. "Cryptozoa." *Science and Children,* March 1987, pp. 34–36.

Miller, G. Tyler. *Environmental Science: An Introduction,* 2nd ed. Belmont, CA: Wadsworth Publishing Co., 1986.

Nadel, Ira Bruce, and Cornelia Hahn Oberlander. *Trees in the City.* New York: Pergamon Press, 1977.

Investigation 9

# Impact of People
# on the Environment

## Student's Section

You already know quite a bit about this subject, so it will begin with a discussion. That's a good way for everyone to share what he or she already knows and to learn something new in the process. In preparation for the discussion, think about these questions and make some notes so you'll have something definite to contribute.

1. What is meant by *impact*? Does the word have only a negative connotation—that is, adverse or detrimental effects—or does it have a positive connotation as well? (Look up *impact* in the dictionary, but also think about how it is used in an environmental sense.)

2. In general, how do people exert an impact on their environment?

3. What are the particular impacts of people on a school environment?

4. List some examples of environmental impact at your school.

5. Which environmental impacts would be most interesting and possible to investigate? How would you go about investigating them? Devise some procedures.

Some of the ideas generated in the discussion may not be possible to implement for a number of reasons, primarily time, expense, and expertise, but you may find ones that can be done.

This investigation has no data sheets, but that doesn't mean there are no data. Devise your own sheets, using the other data sheets as an example. Remember that records are important. You will be discussing the results of your investigation in class. If you don't have good records, you'll have nothing to discuss, and that invalidates your whole investigation.

Investigation 10

# Natural Areas

## Teacher's Section

Many schools are built at the outskirts of their communities in campus-type settings that include considerable acreage. Frequently small natural areas are preserved, either for aesthetic reasons or because alterations, such as the removal of a rock outcrop, would be prohibitively expensive. Seldom is the preservation of these natural areas considered from the educational point of view.

These areas are manipulated to a variable extent. Some are left entirely alone and so are, indeed, natural areas, while others may be altered variously in the interests of safety or health or aesthetics. The most common changes involve water. Low-lying or swampy areas are often drained and filled in, or dredged and shaped to form ponds, while running water is usually channeled with artificial beds. Planting, appropriate or not, is frequently added. Once these areas are set aside or created, they are virtually ignored and become pocket-sized areas of the nearly wild as native species gradually take over.

Such areas are natural biology labs, a fine resource that is all too seldom used. This investigation provides you with some suggestions so you can use such areas more effectively.

The natural areas found on school grounds—woods, field, desert, pond, and stream—reflect the natural ecosystems of that part of the country. Some general information on these ecosystems is included below before discussing specific procedures.

### Terrestrial Ecosystems

Terrestrial ecosystems are of three major types: desert, grasslands, and forest. Plants, the dominant form of life, characterize and often name ecosystems; they create the conditions animals require and are, in turn, affected by climatic and soil conditions. These conditions of precipitation, temperature, and nutrients are limiting factors that determine where particular species can live. For plants, the most important limiting factor is water. The average annual precipitation determines whether desert, grasslands, or forest will develop in an area, while the average annual temperature determines the particular type of ecosystem: cold (polar), moderate (temperate), or hot (tropical). On a worldwide basis there are nine different ecosystems, resulting from the interaction of the three precipitation and three temperature patterns. Those pertaining to the United States and Canada are considered briefly here.

### ● DESERT

The average precipitation is less than 25 cm (10″) per year. Temperate deserts have warm temperatures year-round (southwestern United States), while cold deserts (Great Basin) have cold winters and hot summers. Vegetation includes various shrubs (sagebrush, mesquite), cacti, and small flowering, herbaceous plants; typically the plants are low-growing and well separated from one another.

### ● GRASSLANDS

The average precipitation ranges from 25 to 75 cm (10 to 30″) per year. The temperate grasslands of North America, the ecosystem of the plains in the interior of the continent, have cold, snowy winters and hot, usually dry, summers. Grasses are the dominant vegetation; there are few trees. The fluctuation in precipitation together with natural fires prevent forests from developing even where there is enough water. The cold grasslands (tundra) are found both at high latitude (Arctic) and at high altitude (mountains) and have small shrubs and flowering plants as well as grasses.

### ● FOREST

Trees require a great deal of water, and forests are found wherever the average precipitation exceeds 75 cm (30″) per year. This is the natural ecosystem of much of North America: Alaska, across Canada from coast to coast, the Pacific Northwest, and the eastern half of the United States. Such a large expanse of land covers a wide range of temperatures and consequently includes different types of forest. These are broadly grouped into the northern coniferous forests (spruce, fir, pine), temperate deciduous forests (oak, maple, beech, hickory), and tropical forests (broad-leafed, nondeciduous trees).

## Aquatic Ecosystems

There are several types of aquatic ecosystems, but only two of them are features of school grounds. Here, too, plants are the dominant form of life. The limiting factors of light (pond) and current (stream) determine which kinds of plants can live under the specific local conditions.

### ● PONDS

Ponds are small, basin-shaped bodies of water with sloping shorelines. The water is still, shallow, warm, and murky, the bottom mucky. Rooted plants grow along the shoreline and into the open water (reeds, cattails, water lilies), while algal colonies float near the surface.

### ● STREAMS

Water flowing down a slope forms a stream. If the slope is low, the flow is slow and rooted plants abound. The steeper the slope, the more rapid the flow; rooted plants cannot get established, and algae are the only vegetation.

# *Procedure*

You will need to survey natural areas at your school in order to have an idea of their composition and educational value. The procedures given are general, of necessity, and the survey will provide you with the specifics about your particular area. Don't feel, though, that you have to know everything before taking your students to the area; you can discover it along with your class.

The class should first determine the size of the area and its distance from the building. Since approximate measurements are perfectly adequate for that, you can have several students pace the area off. Measure each one's normal stride, that is, the distance between alternate footfalls, or between the toe of the stationary foot and the touchdown of the heel of the moving foot (usually one to two feet). The pacers then walk around the natural area, maintaining the same stride as best they can, and thus determine the perimeter. To convert that to area, the more meaningful measurement, students make a sketch of the natural area, then square it up as closely as possible and determine the size in square feet (length × width). That figure is then converted to acres or fractions of an acre (1 acre = 43,560 square feet).

The different natural areas are described below, and the aspects that students should observe or determine are indicated on the data sheets.

## 1. Woods

A pocket-sized woods, whether a forest leftover or a once-planted area gone wild, is the most common natural area on school grounds.

The largest and most complex ecosystem in North America is the deciduous forest ecosystem. These forests are stratified vertically into four zones: the canopy, the understory, the shrub, and the herb zones. The canopy, exposed to the sky, is composed of the overlapping branches of the tallest trees. The understory consists of shorter trees, whose tops reach the lower branches of the canopy trees. Shrubs, multi-stemmed woody plants, are still shorter, typically to the level of the understory branches, while the herbaceous plants (ferns are the most common) are at ground level. Most light, obviously, falls on the canopy; as it passes downward it is filtered by the leaves so that lower levels receive progressively less light. Plants live in a particular zone not by accident, but because the amount of light there meets their requirements. Growing plants of different ages (Investigation 3) are present, but very few young trees of the canopy species. They require more light than is available at ground level, so their germination and development await a break in the canopy as a result of blowdown or cutting.

The surface of the ground is covered with litter, mostly decomposing leaves, but a few inches down in the moist soil, decomposers have converted the litter to humus. Most of the soil organisms (Investigation 5) abound, but higher vertebrates are generally limited to rodents and birds. The inhabitants vary with the locale.

Zonation patterns are not a constant feature of forests. Different types of forests (deciduous, coniferous, mixed deciduous-coniferous, and tropical) all have somewhat different patterns. A small woods has more light penetration than a forest and often shows modified zonation.

This self-contained little ecosystem is worth investigation. You can point out some of the salient features, such as zonation, but mostly let students explore on their own. They should identify the dominant trees, discover how they create habitats for other life, see decay and soil formation, look for microhabitats (under logs, rocks, and shrubs), dens (in trees and the ground), nests, and animals (birds, primarily). Soil analysis (Investigation 4) can also be done if you wish. Most importantly, students should see this as a whole community of organisms that interacts together.

## 2. Field

Many schools have unmowed fields of grasses and weeds on the periphery of the landscaped grounds. This is not a reflection of the natural ecosystem, but of openness: any cleared area becomes a field of herbaceous plants. However, in natural grasslands ecosystems, this growth is a permanent feature, whereas it is not in a naturally forested ecosystem; there the herbaceous plants are only temporary, a step in succession, soon replaced by shrubs and trees. Grasses can be maintained in such an area only by mowing, which kills young woody plants.

One of the principal adaptations of grasses and weeds is their capacity to reproduce asexually, usually from the crown. As a result, a field generally becomes a dense mass of vegetation, so dense in fact that it is hard to see the ground. The most common animals are herbivorous insects and their predators.

The field, though ecologically simpler than the woods, is also a self-contained community of interactive organisms. Students should discover the dominant species and observe the close packing of the plants, as well as the animals present. Soil analysis reveals little new information and need not be done.

## 3. Desert

Some school grounds have remnants of natural desert. A fragile ecosystem, its inhabitants are all well-adapted to life in arid environments. The principal adaptations of plants include: the ability to store water (cacti); extensive root systems (mesquite); thorns, spines, or unpleasant taste (mesquite, creosote bush, cacti, sagebrush) that discourage herbivory; and means of decreasing water loss or increasing absorption. Desert plants are separated from one another, an adaptation that reduces competition for resources. Sexual reproduction is limited to favorable environmental conditions, such as rain. Typical animal species include arthropods, reptiles, birds, and mammals, all of which have their own adaptations to desert life, including limiting activity to the cooler times of the day and various means of conserving water.

The little school desert may not be entirely self-contained, depending both on its size and on its proximity to the artificial environment. Community life probably will not be apparent. Students should identify the dominant plant species, noting their spacing and other adaptations, and look for animals and their evidences (chewed vegetation, tracks, feces, dens, feathers, tufts of fur caught on thorns). Soil samples can also be collected for analysis (Investigation 4) and for organisms (Investigation 5) if you wish.

## 4. Pond

Little ponds are natural features at some schools and artificially created at others. Natural ones are spring- or stream-fed and drain, insuring a water turnover that artificial ponds may lack. Such ponds are stagnant, with very little life.

Ecologists divide the pond ecosystem into six zones, each of which is a different habitat, as follows:

### ● *THE EMERGENT ZONE*

This is the water's edge, where plants are rooted underwater with their stems and leaves above water. Cattails, water grasses, sedges, and rushes predominate. Microbes and small animals, such as worms, snails, insects, small fish, and frogs live there, while birds and mammals visit the area.

### ● *THE FLOATING-LEAF ZONE*

Here the water is deeper and the plants, also bottom-rooted, have long underwater stems supporting leaves and flowers that float on the surface. Water lilies are the most common example. A variety of life, both microscopic and macroscopic, lives among the stems, and the area is a breeding ground for fish.

### • *The Submergent Zone*

The water is deeper still, and bottom-rooted plants such as pondweeds are entirely underwater. Microbes, snails, insects (particularly larvae), and fish also inhabit the area.

### • *The Surface-Film Zone*

The water's surface tension is strongest in this thin film. Insects (water striders and whirligig beetles) that merely dent the surface as they walk across it are the most common inhabitants. Algae and floating plants such as duckweed are the usual plants.

### • *The Open-Water Zone*

This variable zone, when present, is confined to the middle of the pond; it does not exist where rooted vegetation extends across the pond. The zone is dominated by plankton (a conglomeration of algae, bacteria, protozoa, rotifers, crustaceans, and larvae) and by insects and fish.

### • *The Bottom Zone*

The organisms of decay, including bottom-living fish, microbes, worms, insect larvae (especially mayfly, caddisfly, and stonefly larvae), and crayfish live in the dark, detritus-covered bottom. Most of them live on top of the muck, some in it, but since the oxygen level decreases with depth, most of the muck cannot support life.

In ecological terms, a pond is impermanent. Vegetation growing in the water and along the shore is the major contributor to its demise. That happens in two ways, from the shore and from the bottom. Shoreline growth restricts the edges, and dead plant materials settle on the bottom where they decompose, adding to the bottom muck; in time, the pond is entirely filled in.

Because a pond is shallow, the water is warm, and warm water holds less oxygen than cold water does. The water is cloudy from muck in suspension, limiting the depth to which light can penetrate and thus confining photosynthetic organisms to the top few inches. Some of the oxygen they produce passes out of the system to the air above; decomposition also uses oxygen, so the amount of dissolved oxygen in a pond is less than it is in other bodies of water. Pond animals are those that have low oxygen requirements, such as the "warm-water fish" (suckers, catfish, sunfish, carp).

The pond ecosystem should be investigated both physically and biologically.

## Physical Properties

The important physical properties are turbidity, depth, temperature, pH, and dissolved oxygen. All are easily determined and can be measured by the class as a whole or by teams. Students should see the physical factors not as independent entities, but as the factors that determine which organisms can live in the pond. They may need your help to make the connection.

Test kits are available from supply companies to determine freshwater pH (neither pH paper nor soil test kits are effective) and dissolved oxygen. While not inexpensive, they are the only means to measure these important variables, and the kits supply more than enough chemicals for the class. If your budget is feeling flush, you can also measure dissolved carbon dioxide, but this is the least necessary test.

### • *Turbidity*

The amount of cloudiness in the water is a measure of the depth to which light can penetrate. The apparatus used to measure it is the Secchi disk, a circular disk divided into alternating black and white quarters and lowered on a calibrated line to determine the depth at which it just fades from view. Photosynthesis, actively occurring at the water's surface, gradually diminishes with depth and ceases before the disk is lost to view.

You can save the cost of a disk with a homemade model: a Frisbee with the quarters painted on it or an aluminum pie pan, unpainted. Make a hole in the center of the disk, put a small ring bolt through the hole, and run stiff cord through it. Test it to be sure it sinks, and add more weight to the nut if need be. Calibrate the line by marking it with a waterproof marking pen at three-inch intervals.

The disk should be lowered straight down with the line at right angles to the water's surface. The easiest way to do that from shore is to attach the calibrated line to a fishing pole. (A pier or bridge is a very useful addition to a school pond; if yours is not so equipped, do try to get one or the other.)

### ● *Depth*

Lower the disk all the way to the bottom and determine the depth at several sites.

### ● *Temperature*

Use the calibrated line and fishing pole to lower a thermometer and record the temperature at different depths: at the surface, in the middle, and at the bottom. The soil thermometer, with its protective shield, is the best instrument for the purpose.

### ● *pH*

Use a meat baster to obtain water samples, then follow the test kit's directions for measuring the pH, probably at several sites.

### ● *Dissolved Oxygen*

Use the same water sample to test for dissolved oxygen, following the test kit's directions.

## Pond Life

Collecting water organisms does require some equipment. A long-handled net, variously called a dip or aquatic or collecting net, is indispensable for larger organisms and can be purchased from supply companies. Other equipment includes a plankton net, meat basters, sieve, pail, shallow white dishes and pans, hand lenses, and forceps.

Plankton nets are much more expensive than dip nets, but can be made quite easily. To do so, you need a metal ring about six inches in diameter, a nylon stocking, a small narrow jar with threaded rim (olive bottle size), string, wire, and duct tape. Mark three equidistant positions on the ring, put the top of the stocking over the ring and secure it in place by twisting the wire tightly around the ring at each of the marked positions. Twist the free ends of the wire together and make a loop to attach a line and then cover all wire ends with duct tape. Cut off the foot of the stocking and tie it to the rim of the jar with string.

Collecting activities at the pond can be done by the class as a whole or by teams and can include any or all of the following:

### ● *Distinguishing Zones*

Have your students locate the emergent, floating-leaf, and submergent zones, identify the dominant plant species in each, and look for animals along the shoreline and on the vegetation. The long-handled net can be used to hook submerged plants as well as to catch animals amid the vegetation. The catch should be put in the pan, with water, examined, identified, and returned to the pond.

### • *WATER-FILM INSECTS*

Students should watch the way these insects walk on the water and try to figure out how they can do so. It is possible because the tendency of the water molecules to cling together (surface tension) is greater than the weight of the insect. The position of the legs, out to the sides rather than under the body, helps to distribute the weight. Students can also determine the weight of an insect to give them an idea of the strength of this film of water. They collect a handful of insects, put them in a container and weigh it, then subtract the weight of the empty container. Dividing the answer by the number of insects gives the weight of a single insect.

### • *OPEN-WATER INHABITANTS*

Use the meat baster to obtain water samples from different areas. You may be able to see some organisms in a white dish, but most of them are too small and require microscopic examination.

The plankton net, which is dragged through the water, is an even better way to obtain samples. The easiest way for shore-based people to operate it is to tie a weight to the end of a rope and throw it across the pond to a waiting catcher. If that isn't possible, have someone walk around the pond, unrolling a ball of string attached to the net's line as he or she goes. In either case, the net winds up being on one side of the pond with its line stretched across to a student on the opposite side who then *slowly* reels it in. If the collecting bottle is held up to the light, some organisms can be seen, but here, too, microscopic examination is needed.

When students examine the samples, they should try to get an idea not only of the diversity of life but also of the proportion of photosynthetic organisms present. These are the organisms that determine the life the pond can sustain.

Larger open-water organisms, both plant and animal, can be caught with the long-handled net. Students should explore the pond systematically by slowly swinging the net through it (another instance where a pier or bridge is handy). After each swing, lift the net and flip it over to prevent escape. The captives are then poured into the pan for examination and rough identification (insects, fish) before being returned to the pond. Interested students can pursue that further by identifying the species and learning more about them.

### • *BOTTOM LIFE*

The mucky bottom is full of assorted nonphotosynthetic life. Some animals (crayfish, insect larvae, fish, turtles) hide in crevices between rocks and can be discovered by tipping rocks over with the long-handled net. Other bottom life will be found clinging to detritus, or anchored on rocks, or living in the muck. Organisms include microbes, sponges, worms, mollusks, and insects.

If possible, collect samples from the surface of the muck as well as down in it. A borrowed pair of boots or fisherman's waders will enable one member of the class to get out into the pond to collect. He or she needs only two pieces of equipment: a plastic pail (child's beach toy or painter's pail) and a small kitchen sieve. Muck scooped up with the sieve is collected in the pail; back on shore, it is poured into the pan to examine. That should include its composition (Investigation 4) as well as the organisms. This is a good occasion to make the physical determinations of the muck (temperature, pH, dissolved oxygen).

## 5. Stream

Streams, or flowing water, are either natural or artificial features. In a school setting, natural streams are usually found in conjunction with a pond, as its inflow and/or out-flow, and tend to be fairly slow-moving. Drainage ditches, found on many school grounds, are artificial streams, but like natural ones have their typical populations and make worthwhile subjects for study.

Current speed is the main factor determining the life of the stream, and it depends on the steepness of the slope. If the slope is so low that the water barely moves, then it is a slow stream, virtually an elongate pond. Treat it as such. Streams with steeper slopes are altogether different. In these swift streams, water speed is variable, cascading over rocks and flowing more slowly in basins, forming two quite different habitats—rapids and pools—in close proximity.

The speed of the current is measured by determining how long it takes an object to travel a given distance. Choose a typical section of the stream, about 25 feet long, and have students tie a marker flag onto stream-side plants at each end and measure the distance between the flags. A student who has a watch with a second hand is then stationed at each marker. The person at the upstream end tosses a stick or small piece of wood (2 to 4 inches long) into the current while noting the time; the student at the downstream end notes the time when the stick passes the second marker. Current speed is the distance divided by the time and is expressed as feet per second; for example, if the stick traveled 25 feet in 30 seconds, then the current speed is 25/30 or 0.8 feet per second. For accuracy, the procedure should be repeated three times and the average figured. Speeds in different sections can also be determined.

Fast-moving water is clearer, colder, and more oxygenated than slow-moving water. Tumbling water sends find spray aloft, where it becomes saturated with air, whereas standing water loses oxygen to the air.

If oxygen content were the governing factor, you would expect a stream to sustain more life than a pond, but this is not the case; current force is the stream's limiting factor. It prevents rooted plants from getting established, restricting plant life to algae and limiting animals to those with adaptations that prevent them from being swept away (living in rock crannies, burrowing in the bottom, construction of tubes and cases, suc-tion devices, and anchoring cement). Current also tends to remove detritus, which is another limiting factor.

In addition to current speed, students should determine other physical properties of the stream: clarity or turbidity, depth, temperature, pH, and dissolved oxygen content; the procedures are the same as those given for the pond.

The life of the stream varies with habitat (rapids or pools). Animals are more com-mon than plants and include worms, snails, crustaceans, insect larvae (especially mayfly, caddisfly, and stonefly larvae), fish, and salamanders; they feed on detritus or on one another. Species differ with the habitat. Rooted plants are confined to stream edges, a well-inhabited area, and algae are found in flowing water. Channel life is not readily apparent at first glance. Fish can be seen as long as the class is quiet and the water clear, but most life exists beneath rocks. Have students pick up rocks and examine their under-sides, where the students will find adhering insect larvae along with colonies of gelati-nous algae; crustaceans and salamanders will scurry for shelter when the rocks are removed. The dip net or aquarium net or small sieve can be used to catch these crea-tures for examination, which should be done quickly, with little handling. The animals should be returned to the stream and the rocks replaced as they were. Students can also take water samples at different locations and depths to examine for microbial life.

Examining the stream and its life can be done by the class as a whole, or you can assign different areas to different teams.

# *Discussion*

This investigation is entirely observational, but it is important for students to have the opportunity to discuss the significance of their findings afterward. Some of the salient points that need to come out in the discussion are included in the comments after the questions below.

1. Ecosystems have a few plant species that are said to be "dominant." What does "dominant" mean in that context?

   *Comment:* Ecologically, the term means predominant in the sense of influence, not necessarily numbers. The dominant plants are those that create the ecosystem: canopy trees in a woods, grasses in a field, shrubs in a cold desert, cacti in a warm one, algae in open or flowing water, flowering plants in still, shallow water.

2. How does the community sustain itself? Do you think it is large enough to be entirely self-sufficient? If not, what comes in from the outside?

   *Comment:* Students should arrive at the idea of mutual dependency, which includes but goes beyond the food web. Probably the best word, and one students will surely find, is *home*. The natural area is home to its residents; it provides them with shelter and whatever semblance of security nature offers as well as the necessities of food, water, oxygen, carbon dioxide, and soil. Another important point about interdependency is recycling: absolutely nothing is wasted. Whether or not the community is entirely self-sufficient depends on its size; the larger it is, the more likely it is able to sustain itself. It is not a closed community—none is—and animals that wander through it leave seed-containing wastes that may later germinate and become part of the community. Other wanderers may remain.

3. How much do the activities of the school affect the life in these natural areas?

   *Comment:* That depends on the proximity of the building to the area. Noise and trampling of soil and vegetation are the major effects (Investigation 9).

4. What makes pond water so cloudy? If you could get rid of the cloudiness and have perfectly clear water, what would happen to the pond life?

   *Comment:* Turbidity results from materials in suspension and its origin is biological. Muck is the watery equivalent of humus (minerals and decomposed organic material). Ponds have a great deal of life and a great deal of muck; this is stirred up by the activities of the animals, and consequently the water is cloudy. Some ponds are cleaned by dredging and filling with fresh water. This accomplishes nothing in the long run because the conditions that favor a pond ecosystem still exist, and the surviving life soon reconstitutes it.

5. How do the physical factors of size, depth, water movement, turbidity, temperature, pH, and dissolved oxygen collectively determine the kinds of life that can live in the pond?

   *Comment:* The pond, a basin of murky, warm water with low oxygen content and a pH that reflects the surrounding geology and vegetation, is the natural habitat of organisms that require those conditions. That's their home. It selects them initially, but their activities tend to advance the very conditions that favor them, and so they gradually alter conditions to the point where they can no longer survive (acidophilic plants make the water more acidic, for example, even to the point where it becomes too acidic for them). As conditions change, other organisms gradually become established and eventually take over the pond.

6. How does a stream's current affect the organisms living in the stream?

   *Comment:* The current sweeps away detritus, preventing its decomposition and the formation of bottom muck that supplies nutrients for life. It also removes organisms unable to resist the flow so that the only organisms that can live in the swift stream are those adapted to its conditions.

7. What adaptations enable organisms to live in a particular ecosystem?

   *Comment:* Students should get beyond the obvious—anatomical adaptations—and consider the physiological ones that are responses to the particular conditions. These range, for example, from water use efficiency in desert dwellers to means of excluding water in aquatic ones, from photosynthesis in a dim woods or pond to photosynthesis in the bright light of an open field.

   Try to guide student ideas toward an understanding of some basic biological facts: that organisms live where they do because they can—the conditions there meet their requirements—and that organisms are not perfect but a collection of compromises that work superlatively well under those conditions. That is true for all life, plant and animal and microbe alike.

8. If you found a situation in which there were very few organisms, what might be responsible?

   *Comment:* Various environmental conditions, notably noise (higher animals) and runoff pollution (pond or stream), could account for a scarcity of organisms, but superficial examination by students is probably primarily responsible. Students tend to miss a great deal. Life is out there to be found, if one will but look.

9. What differences did you notice between the natural areas and the landscaped areas? What causes these differences?

   *Comment:* While answers depend to a large extent on the particular situation, there are some common features. Landscaped areas are artificial ones that must be maintained or they will return to the wild. So lawns are mowed, shrubs trimmed, and undesirable plants removed. Many landscape plants are not native, are less vigorous in that environment than native ones are, and require more care (which they may or may not receive). Higher animals tend to be absent, save for opportunistic species. There is little diversity of life. All those attributes contrast sharply with natural areas in which life is rampant, diversity abounds, every niche is filled, and keen competition for the necessities of life holds populations in check.

   Students should compare the managed and unmanaged worlds, determine which is more stable and why (diversity and stability go together for reasons we barely perceive), and compare the results of previous investigations (especially 2, 3, 6, 8, and 9) with this one.

## Supplies for Terrestrial Areas

| | |
|---|---|
| guides (see general reference list, pages *xiv–xvii*) | hand lenses<br>baby food jars with masking tape labeling strips |

## Supplies for Aquatic Areas

| | |
|---|---|
| guides (see general reference list, pages *xiv–xvii*)<br>Secchi disk, or Frisbee or aluminum pie pan with ring bolt and stiff cord<br>calibrated line<br>soil thermometer<br>freshwater pH test kit<br>freshwater dissolved oxygen test kit<br>long-handled collecting net<br>plankton net<br>rope and weight or ball of string<br>fishing pole<br>meat basters<br>kitchen sieve<br>small pail | aquarium nets<br>white dishes and pans<br>hand lenses<br>forceps<br>baby food jars with masking tape labeling strips<br>boots or waders (optional)<br>steel tape, 8′ or longer<br>marker flags<br>watches<br>yardstick<br>diet scale<br>microscopes<br>slides and droppers |

## Spinoff Idea

Suggestion for a spinoff investigation:

- How Do People Investigating a Natural Area Affect It?

## References

Allen, Durward L. *The Life of Prairies and Plains.* Our Living World of Nature Series. New York: McGraw-Hill, 1967.

Amos, William H. *The Life of the Pond.* Our Living World of Nature Series. New York: McGraw-Hill, 1967.

Amos, William H. *Limnology.* Chestertown, MD: LaMotte Chemical Products Co., 1969.

Baumann, Richard W. "Water Insects and Their Relatives." *American Biology Teacher,* May 1977, pp. 295–298.

Botkin, Daniel B., and Edward A. Keller. *Environmental Studies: The Earth As a Living Planet.* Columbus, OH: Charles E. Merrill Publishing Co., 1982.

## *References (continued)*

Brown, Lauren. *Grasslands.* Audubon Society Nature Guides. New York: Alfred A. Knopf, 1985.

Caduto, Michael J. *Pond and Brook: A Guide to Nature Study in Freshwater Environments.* Englewood Cliffs, NJ: Prentice-Hall (date unavailable).

Garber, Steven. *The Urban Naturalist.* New York: John Wiley and Sons, 1987.

Headstrom, Richard. *Suburban Wildlife.* Englewood Cliffs, NJ: Prentice-Hall, 1984.

Krall, Florence. "Mudhole Ecology." *American Biology Teacher,* September 1970, pp. 351–353.

MacMahon, James A. *Deserts.* Audubon Society Nature Guides. New York: Alfred A. Knopf, 1987.

Mattingly, Rosanna L. "Turning Over a Wet Leaf." *Science Teacher,* September 1985. pp. 20–24.

McCormick, Jack. *The Life of the Forest.* Our Living World of Nature Series. New York: McGraw-Hill, 1966.

Miller, G. Tyler. *Environmental Science: An Introduction,* 2nd ed. Belmont, CA: Wadsworth Publishing Co., 1986.

Line, Les, and Lorus and Margery Milne. *Audubon Society Book of Insects.* New York: H.N. Abrams, 1983.

Milne, Lorus, and Margery Milne. *Invertebrates of North America.* New York: Doubleday (undated).

Phillips, Roger E., Jr. "A Field Trip to the Stream," *Carolina Tips,* February 1984. Carolina Biological Supply Co., Burlington, NC 27215.

Rushforth, Samuel R. "The Study of Algae." *American Biology Teacher,* May 1977, pp. 316–320.

Slavik, Bohumil. *Wildflowers: A Color Guide to Familiar Wildflowers, Ferns and Grasses.* London: Octopus Books, 1973.

Smith, Howard G. *Tracking the Unearthly Creatures of Marsh and Pond.* Nashville, NY: Abingdon Press, 1972.

Sutton, Ann, and Myron Sutton. *Eastern Forests.* Audubon Society Nature Guides. New York: Alfred A. Knopf, 1988.

Sutton, Ann, and Myron Sutton. *The Life of the Desert.* Our Living World of Nature Series. New York: McGraw-Hill, 1966.

Usinger, Robert L. *The Life of Rivers and Streams.* Our Living World of Nature Series. New York: McGraw-Hill, 1967.

Whitney, Stephen. *Western Forests.* Audubon Society Nature Guides. New York: Alfred A. Knopf, 1985.

Winget, Robert N. "Aquatic Life in a Utah Desert." *American Biology Teacher,* May 1977, pp. 279–284.

## Investigation 10
# Natural Areas

## Student's Section

In this investigation, you'll be examining natural areas on the school grounds, places that have been left on their own without much human interference. There's no lawn or mowing, no planted and trimmed shrubbery. It's rough, not neat.

These areas may be remnants of native ecosystems or an unmaintained area gone wild. As you investigate, you'll be able to see what natural ecosystems in your part of the country look like and appreciate how much the environment has been changed to create the grounds surrounding the school.

You'll need to know about how large the area is. The easiest way to do that is to pace it off. A person's stride is the distance covered by alternate footfalls, or from the toe of the stationary foot to the touchdown of the heel of the moving foot. The length depends on a person's height and speed, but one to two feet is about average.

Several people in the class can walk around the area, counting their paces as they go and tallying the results afterward. The method is not exact, but you do want to get as good an approximation as possible, so first measure the pacers' strides, and then have the pacers try to maintain the same stride all the way around the perimeter.

On the data sheet, make a sketch of the area's shape, including the length of each "leg" or section, as well as the whole perimeter. After you have done that, square off the sketch and determine the area, first in square feet (length × width) and then in acres or parts of an acre (1 acre = 43,560 square feet).

Record your other observations on the data sheet.

*Biology Is Outdoors!*

# The Pond

Plants create the pond community; they are the most obvious organisms, they are the producers of food and oxygen, and they make the habitat for all other living things. Ecologists classify the pond ecosystem into six zones, based on the growth habits of the dominant plants in each zone. These are:

### 1. THE EMERGENT ZONE

This is the water's edge where plants are rooted underwater, with their stems and leaves above water. Cattails, water grasses, sedges, and rushes are typical. Small animals, such as worms, snails, insects, small fish, and frogs live there, while birds and mammals visit the area. There are many kinds of microbes as well.

### 2. THE FLOATING-LEAF ZONE

Where the bottom slopes down and the water is deeper, the plants, also bottom-rooted, have long underwater stems supporting leaves and flowers that float on the surface. Water lilies are the most common example. There are a variety of microbes and animals, and the area is a breeding ground for fish.

### 3. THE SUBMERGENT ZONE

The water is still deeper, and bottom-rooted plants such as pondweeds are entirely underwater. Microbes, snails, insects (particularly larvae), and fish also inhabit the area.

### 4. THE SURFACE-FILM ZONE

This is the narrowest zone, right at the surface of the water, inhabited mostly by water striders and whirligig beetles, insects that can walk on the water. Algae and floating plants such as duckweed are also found there.

### 5. THE OPEN-WATER ZONE

This zone is confined to the middle of the pond, or it may not even exist if rooted vegetation extends all the way across the pond. It is dominated by plankton (a mixture of algae, bacteria, protozoa, rotifers, crustaceans, and larvae) and by insects and fish.

### 6. THE BOTTOM ZONE

The bottom is dark and mucky and inhabited by bottom-living fish and organisms of decay such as microbes, worms, larvae, and crayfish. Most of them live on top of the muck, but a few can live in it.

# *Directions for Obtaining the Data*

## Physical Properties

First, you'll be doing some tests to determine the pond's important physical properties. Try to understand how these factors can determine the life in the pond.

### *1. TURBIDITY*

The amount of cloudiness in the water is a meaure of the depth to which light can penetrate and photosynthesis occur.

Lower the Secchi disk slowly into the water, keeping the line perpendicular, and record the depth when the disk just fades from sight.

### *2. DEPTH*

Lower the disk all the way to the bottom and read the depth from the line. Do this at several sites.

### *3. TEMPERATURE*

Attach the thermometer to the calibrated line and measure the temperature at several depths: at the surface, in the middle, and at the bottom. Don't be too fast to pull the thermometer up, since it does take a few minutes for the mercury to reach temperature.

### *4. pH*

Collect water samples with the baster. Put each sample in a separate, labeled jar, then follow the test kit's directions for determining pH.

### *5. DISSOLVED OXYGEN*

Use the same water samples to test for dissolved oxygen, following the test kit's directions.

## Pond Life

1. Locate each of the zones that is present (some may not be) and identify the dominant plant species in each one. That will be harder to do in some zones than in others, and you may need to snag plants with the long-handled net. Look for animals and try to catch some with the net. Put them in a pail or pan of water, examine them quickly, and return them to the pond.

2. Watch the way insects walk on the water and try to figure out how they can do that. Catch a few and put them in a pan of water so you can watch them more closely. Notice the position of their legs and how their weight is distributed. Find out how much an insect weighs. Collect a bunch of them in a container, count and weigh the group, empty them back into the pond, and weigh the empty container.

Subtract the weight of the container from the weight of the container plus insects and divide the answer by the number of insects to find the weight of one insect. That gives you an idea of the strength of the water's surface film.

3. Use the baster to collect water samples and put each in a separate labeled container. You may be able to see some organisms, but microscopic examination is necessary for most.

4. The plankton net is a better way to collect specimens from the parts of the pond that you can't reach. As it is *slowly* pulled across the pond, water and organisms fill the collecting bottle. Untie the bottle and hold it up to the light, and you may be able to see some of them. Remove clinging organisms from the net. Examine the organisms you can see and take samples back to the lab for microscopic examination. Make several collection runs with the net if you wish.

   When you examine water samples microscopically, look at the variety of living things, but concentrate mostly on the proportion of photosynthetic (green) organisms that form the base of the food web.

5. Collect larger organisms, both plant and animal, from the open water with the long-handled net. You can go after fish, but don't ignore the smaller organisms, ones that you may not even be aware of. Swing the net *slowly* through the water in a systematic fashion, lifting the net after each swing and flipping it over to prevent captive animals from escaping. Put your catch in a pan or pail of water to examine, and return the organisms to the pond afterward.

6. The mucky bottom is full of life. Larger animals often hide in crevices under rocks, and you can find them by tipping the rocks over. If you use the long-handled net for that, you may be able to tip and collect in one maneuver. Other bottom-living organisms will be found clinging to detritus (wastes, such as leaves and sticks), or anchored on rocks, or living in the muck.

   Collect samples at the surface of the muck as well as down in it. Wade out into the pond and scoop up muck with the sieve and put it in the pail. Back on shore, pour the muck into the pan to examine. Look at its composition, as you did with the soil in Investigation 4, as well as the organisms, and measure the physical properties of temperature, pH, and dissolved oxygen. Record your results on the data sheet and return the muck to the pond when you've finished.

# The Stream

The stream you're going to be investigating may be a natural or an artificial one. In either case, the speed of the current determines whether it is a slow or a fast stream, and that determines which kinds of organisms can live there. Notice the way the channel is formed, with plants growing along the edges and rocks usually lying in it. If the slope is steep, the water moves fast, tumbling over the rocks in rapids; in other places the rocks form basins, deep pools of barely moving water. These are different habitats with different kinds of life.

# Directions for Obtaining the Data

## Physical Properties

### 1. CURRENT

This is the most important fact of life for stream-dwelling organisms. You can determine the stream's speed by measuring how fast an object travels a known distance between two fixed points. Select a typical section of the stream about 25 feet long, tie marker flags onto stream-side plants at each end of the section, and measure the distance. One person stands at each end of the course. The person at the upstream end tosses a small piece of wood into the current while noting the time. The person at the other end notes the time when the piece of wood passes the second flag. Current speed is the distance traveled divided by the time and expressed as feet per second. For example, if the piece of wood traveled 25 feet in 30 seconds, then the speed of the current is 25/30 or 0.8 feet per second. For accuracy, repeat the procedure three times and determine the average. Speeds in different sections can also be determined.

### 2. CLARITY

Look for fish. If you can see them or the bottom, the stream is clear; if not, it is murky. In that case, lower the Secchi disk into the water with the line perpendicular to the surface and record the depth at which it just fades from sight.

### 3. DEPTH

Clarity is also related to depth; it is easier to see the bottom where the stream is shallow and harder to do so in deeper pools. Use the yardstick to measure the depth at several places.

### 4. TEMPERATURE

Attach a thermometer to the calibrated line and measure the temperature at several depths: at the surface, in the middle, and at the bottom. Don't forget that it does take a few minutes for the mercury to reach temperature.

*Biology Is Outdoors!*

### 5. pH

Collect water samples with the baster. Put each sample in a separate, labeled jar, then follow the test kit's directions for determining the pH.

### 6. Dissolved Oxygen

Use the water samples to test for dissolved oxygen, following the test kit's directions.

## Stream Life

Many organisms live burrowed in the bottom, or in crevices under rocks, or attached to rocks or debris. Pick up rocks and look for organisms on the undersides of the rocks. Watch for scurrying crustaceans and salamanders when you lift the rocks, and try to catch them with the net or sieve. Put them in a pan of water to examine—do that quickly and carefully—and then return them to the stream. Put the rocks back as you found them. Use the baster to collect water samples from different habitats and put them in separated, labeled jars to take back to the lab for microscopic examination. Look at the variety of life present in the stream and at adaptatations to the stream environment, and try to figure out the interactions among the organisms.

Record all your results on the data sheet.

# Investigation 10:
## Natural Areas

| **DATA SHEET**—Woods |
|---|
| Date: |
| Sketch of Area: |

Approximate Distance from Building: _____

Approximate Size of Woods:

   *Perimeter:* _____

   *Area:* _____

     Square Feet: _____

     Acres: _____

Distinctive Physical Features:

   *Rocks:* _____

   *Slope:* _____

   *Gully:* _____

   *Other:* _____

*(continued)*

Name _____ Date _____

# Investigation 10:
# **Natural Areas**

| | |
|---|---|
| **DATA SHEET**—Woods *(continued)* | |

Type of Woods: _____

  *Coniferous:* _____

  *Deciduous:* _____

  *Coniferous-Deciduous:* _____

  *Tropical:* _____

Stratification: _____

  *Zones Present:* _____

Dominant Tree Species: _____

Other Types of Plants: _____

Amount of Light at Ground Level:
          Bright          Speckled       Dim

(Circle the term that most accurately describes the light conditions, or put a check mark on the line where it belongs between two terms.)

Animals: _____

  *Seen:* _____

  *Evidences:* _____

Microhabitats: _____

  *Locations:* _____

  *Inhabitants:* _____

Decay: _____

  *Specimens Decaying:* _____

  *Decomposer Organisms:* _____

  *Seen:* _____

  *Inferred:* _____

Soil Formation: _____

  *Litter:* _____

  *Humus (amount and appearance):* _____

Adaptations to Forest Life: _____


Interactions Within the Forest Community: _____

*Biology Is Outdoors!*

# Investigation 10:
## Natural Areas

| DATA SHEET—Field |
|---|
| Date: |
| Sketch of Area: |
| Approximate Distance from Building: |
| Approximate Size of Field: |
| *Perimeter:* |
| *Area:* |
| Square Feet: |
| Acres: |
| Distinctive Physical Features: |

*(continued)*

# Investigation 10:
# **Natural Areas**

| **DATA SHEET**—Field *(continued)* |
| --- |
| Dominant Plant Species: |
| Other Types of Plants: |
| Amount of Light at Ground Level: Bright _____ Speckled _____ Dim _____ (Circle the term that most accurately describes the light conditions, or put a check mark on the line where it belongs between two terms.) |
| Animals: *Seen:* *Evidences:* |
| Microhabitats: *Locations:* *Inhabitants:* |
| Decay: *Specimens Decaying:* *Decomposer Organisms:* Seen: Inferred: |
| Soil Formation: *Litter:* *Humus (amount and appearance):* |
| Adaptations to Field Life: |
| Interactions Within the Field Community: |

# Investigation 10:
# Natural Areas

| **DATA SHEET**—Desert |
|---|
| Date: |
| Sketch of Area: |
| Approximate Distance from Building: |
| Approximate Size of Field: |
| *Perimeter:* |
| *Area:* |
| Square Feet: |
| Acres: |
| Distinctive Physical Features: |

(continued)

# Investigation 10:
# **Natural Areas**

| **DATA SHEET**—Desert *(continued)* |
|---|

Dominant Plant Species:

Other Types of Plants:

Amount of Light at Ground Level:
       <u> Bright </u>         Speckled <u>    </u>    <u> Dim </u>

(Circle the term that most accurately describes the light conditions, or put a check mark on the line where it belongs between two terms.)

Animals:

*Seen:*

*Evidences:*

Microhabitats:

*Locations:*

*Inhabitants:*

Decay:

*Specimens Decaying:*

*Decomposer Organisms:*

   Seen:

   Inferred:

Soil Formation:

*Litter:*

*Humus (amount and appearance):*

Adaptations to Field Life:

Interactions Within the Field Community:

*Biology Is Outdoors!*

Name _____ Date _____

# Investigation 10:
## Natural Areas

| DATA SHEET—Pond |
|---|
| Date: |
| Sketch of Pond: |
| Approximate Distance from Building: |
| Approximate Size of Pond: |
| *Perimeter:* |
| *Area:* |
| Square Feet: |
| Acres: |
| Source of Water: |
| Outflow: |
| Amount of Water Movement: |

*(continued)*

   *Biology Is Outdoors!*

Name _____  Date _____

# Investigation 10:
## Natural Areas

| DATA SHEET—Pond *(continued)* |
|---|

**Turbidity:** _____

    *Appearance:*      Clear _____      Murky _____

    *Depth at Which Disk Disappears:* _____

    *Where Determined:*

| Depth: | pH: | Dissolved Oxygen: |
|---|---|---|

**Temperature:** _____

    *Surface:* _____

    *Middle:* _____

    *Bottom:*

| pH: | Dissolved Oxygen: |
|---|---|

**Pond Life (Delete zones not present):**

| Zone | Plants | Microbes | Animals |
|---|---|---|---|
| Emergent | | | |
| Floating Leaf | | | |
| Submergent | | | |
| Surface Film | | | |
| Open Water | | | |
| Bottom | | | |

**Proportion of Microscopic Photosynthetic Organisms:**

     None      Few      Many      Most

**Adaptations to Pond Life:**

**Interactions Within the Pond Community:**

(For "Turbidity" and "Photosynthetic Organisms," circle the term that most accurately describes the condition, or put a check mark on the line where it belongs between two terms.)

*Biology Is Outdoors!*

## Investigation 10:
# Natural Areas

| DATA SHEET—Stream |
|---|
| Date: |
| Sketch of Stream: |

Approximate Distance from Building:

Approximate Size of Stream:

*Length:*

*Width:*

Source of Water:

Natural or Artificial:

Current Speed:

*1.*

*2.*

*3.*

*Average:*

(continued

# Investigation 10:
# Natural Areas

| **DATA SHEET**—Stream *(continued)* |
|---|

**Clarity:**

Clear _____ Murky

(Circle the term that most accurately describes the clarity condition, or put a check mark on the line where it belongs between two terms.)

**Depth at Which Disk Disappears:**

**Stream Depth:**

**Temperature:**

*Surface:*

*Middle:*

*Bottom:*

| pH: | Dissolved Oxygen: |
|---|---|

**Stream Life:**

| Habitat | Plants | Microbes | Animals |
|---|---|---|---|
| Rapids | | | |
| Pools | | | |

**Adaptations to Stream Life:**

**Interactions Within the Stream Community:**

© 1991 J. Weston Walch, Publisher

*Biology Is Outdoors!*

# *Spinoff Idea*

## • How Do People Investigating A Natural Area Affect It?

This is a problem ecologists deal with all the time. In fact, scientists in any field know that they themselves affect their studies, often in ways that are difficult to discern. When you devise an experiment, for example, you have a pretty good idea of the results you expect, and it is easy to let your ideas influence the outcome—not intentionally, of course. Scientists need to be objective, but since no one can be entirely objective, it is something scientists have to be aware of and keep working at. So do you.

In some fields, objectivity is particularly difficult to achieve, and ecology is one of them. The living world is made up of many complex interactions that we barely understand. Since we don't understand them, we can hardly control conditions the way we can in a lab. So any ecological study has lots of unknown factors.

In this case, the little natural area at your school has been going along on its own for some time when suddenly it is invaded by a mob of people—your class—intent on studying it. You'd expect some sort of reaction.

As a result of Investigation 9, you are aware of some of the kinds of effects people have on the environment. There are others as well. A few suggestions:

## • Physical Disruptions

Several environmental disruptions occurred in the course of this investigation, such as: turning over stones or logs to look for life underneath, pulling up pondweeds, moving rocks in a pond or stream, walking through a field of high grasses, throwing muck back into a pond. What were the effects? Choose some disruptions that you noticed and try to find out the consequences.

## • Physiological or Psychological Disturbances

The presence of people in the area is the prime example of this type of disturbance.

Vertebrates have acute senses and are wary as a matter of necessity. Their very survival depends on it. Not only do they hear and see any sort of activity in their vicinity, but also other senses come into play, primarily smell, heat detection (ability to sense the presence of a body by the heat it gives off), and vibratory sense (awareness of the presence of a body by the impact it makes on the ground or by the water motion it creates). These animals are aware of your presence long before you notice them, so they hide at the first hint of you. That's why you so seldom see vertebrates in the wild.

As you probably know, scientists who study wild mammals spend a lot of time letting the animals get used to their presence before undertaking any studies. Even though you don't have that kind of time, you can still investigate these animals.

Return to the natural area alone. Look for animals, or places where they congregate, or their dens or burrows. Settle down comfortably with a notebook and make notes about any activities that you see. Compare what you see now with what you saw when the class visited the area. Read about the natural ecosystem in your part of the country so you'll know what kinds of animals you might expect to find and learn something about their habits.

Whatever investigation you undertake, you will need to draw some conclusions afterward. How much of an impact do people who study an area create? Are the effects only temporary, or are there some long-term ones as well? How could people study an area without disturbing it, or with only minimal disturbances?

## *References*

Brown, Lauren. *Grasslands.* Audubon Society Nature Guides. New York: Alfred A. Knopf, 1985.

Grier, James. *Biology of Animal Behavior,* 2nd ed. St. Louis: Mosby, 1989.

Heintzelman, Donald S. *The Birdwatcher's Activity Book.* Harrisburg, PA: Stackpole Books, 1983.

Hickey, Joseph J. *Guide to Bird Watching.* New York: Dover Publications, 1975.

MacMahon, James A. *Deserts.* Audubon Society Nature Guides. New York: Alfred A. Knopf, 1987.

Miller, G. Tyler. *Environmental Science: An Introduction,* 2nd ed. Belmont, CA: Wadsworth Publishing Co., 1986.

Mitchell, John, and Massachusetts Audubon Society. *The Curious Naturalist.* Englewood Cliffs, NJ: Prentice-Hall, 1980.

Nadel, Ira Bruce, and Cornelia Hahn Oberlander. *Trees in the City.* New York: Pergamon Press, 1977.

Sparks, John. *Bird Behavior.* New York: Bantam Books, 1970.

Sutton, Ann, and Myron Sutton. *Eastern Forests.* Audubon Society Nature Guides. New York: Alfred A. Knopf, 1988.

Whitney, Stephen. *Western Forests.* Audubon Society Nature Guides. New York: Alfred A. Knopf, 1985.

Articles on the subject can also be found in current magazines, such as:

*Audubon*          *Science Digest*
*National Geographic*   *Science News*
*National Wildlife*     *Time*
*Newsweek*

# Index

*Note: Boldfaced numbers indicate entries that appear on reproducible student pages.*

## Z